PROCEEDINGS

SPIE—The International Society for Optical Engineering

Human Detection and Positive Identification: Methods and Technologies

Lisa A. Alyea
David E. Hoglund
Chairs/Editors

19, 21 November 1996
Boston, Massachusetts

Sponsored by
SPIE—The International Society for Optical Engineering

Cosponsored by
National Institute of Standards and Technology
ONDCP—Office of National Drug Control Policy
Oak Ridge National Laboratory
Sandia National Laboratories
SAIC—Science Applications International Corporation
Harris Corporation
MITRE Corporation
Lawrence Livermore National Laboratory
Idaho National Engineering Laboratory

Cooperating Organizations
National Institute of Justice
U.S. Customs
Federal Bureau of Investigation
ASPRS—American Society for Photogrammetry and Remote Sensing
CPOA—California Peace Officers' Association
National Forensic Science Technology Center

Volume 2932

Published by
SPIE—The International Society for Optical Engineering

SPIE is an international technical society dedicated to advancing engineering and scientific applications of optical, photonic, imaging, electronic, and optoelectronic technologies.

The papers appearing in this book comprise the proceedings of the meeting mentioned on the cover and title page. They reflect the authors' opinions and are published as presented and without change, in the interests of timely dissemination. Their inclusion in this publication does not necessarily constitute endorsement by the editors or by SPIE.

Please use the following format to cite material from this book:
Author(s), "Title of paper," in *Human Detection and Positive Identification: Methods and Technologies,* Lisa A. Alyea, David E. Hoglund, Editors, Proc. SPIE 2932, page numbers (1997).

Library of Congress Catalog Card No. 96-69895
ISBN 0-8194-2334-3

Published by
SPIE—The International Society for Optical Engineering
P.O. Box 10, Bellingham, Washington 98227-0010 USA
Telephone 360/676-3290 (Pacific Time) • Fax 360/647-1445

Printed in the United States of America.

Contents

Conference Committees

Part A Human Inspection Methods and Technologies: Workplace and Border Crossings

Conference Chair

David E. Hoglund, U.S. Customs Service

Program Committee

Lorne Elias, National Research Council Canada
Joseph J. Fortuna, Chemical Detection Services, Inc.
Terry Gough, University of Leeds (UK)
Susan F. Hallowell, FAA Technical Center
Kevin M. Jackson, National Institute of Justice
Victor Pocaro, U.S. Customs Service

Session Chairs

1 Detection of Contraband Handling
Susan F. Hallowell, FAA Technical Center

2 Detection of Drug Usage
Kevin M. Jackson, National Institute of Justice

3 Detection of Concealed Contraband
Lorne Elias, National Research Council Canada

Part B Positive Identification: Testing Methods and Technologies

Conference Chair

Lisa A. Alyea, Department of Defense

Program Committee

Mark Jones, Lawrence Livermore National Laboratory

Session Chairs

4 Positive Identification: Trends and Technologies
Jay Smart, Lawrence Livermore National Laboratory

5 Biometrics: Trends and Technologies
Lisa A. Alyea, Department of Defense

Part A

HUMAN INSPECTION METHODS AND TECHNOLOGIES: WORKPLACE AND BORDER CROSSINGS

SESSION 1

Detection of Contraband Handling

Cocaine Phenomenology Study
Results of a Third in a Series of Field Trials

Chih-Wu Su
U.S. Coast Guard Research and Development Center
1082 Shennecossett Road
Groton, CT 06340

Steve Rigdon and Steve Ricard
Analysis & Technology, Inc.
258 Bank Street
New London, CT 06320

D. Hoglund, U.S. Customs Service Applied Science Division
G. Drolet, P. Neudorfl, M. Hupe, Revenue Canada
T. Kunz, S. Ulvick, Houston Advanced Research Center
J. Demirgian, Argonne National Laboratory
P. Shier, DynCorp
J. Wingo, Veda

[The opinions or assertions contained herein are the private ones of the authors and are not to be construed as official or reflecting the views of the Commandant or the Coast Guard at large. Any instrument, material or chemical referred to by its brand name does not constitute an endorsement by the authors, the Commandant or the Coast Guard at large.]

ABSTRACT

To form an understanding of the environment in which non-intrusive detection and inspection technologies are required to operate, the Narcotic Detection Technology Assessment Team has undertaken a series of field studies. These field studies have focused on the phenomenology, fate and behavior of narcotic residue in real world environments. The overall goal of the tests is to give Law Enforcement officers the ability to accurately differentiate between individuals involved in the smuggling process and individuals innocently contaminated with narcotics. The latest field study in this series was conducted in Miami, FL in February 1996. The field study comprised several individual tests. The first was a Contamination and Transfer Study which focused on human contamination resulting from contact with actual kilos of cocaine and the mechanism by which this contamination transfers to surrounding objects, if at all. The second was a Secondary Contamination Study which focused on determining the conditions under which cocaine contamination transfers from objects touched by individuals who handled narcotics to innocent passerbys. The third was a Persistence Study which focused on the persistence of cocaine contamination on people under a variety of conditions. An overview of the tests and their preliminary results will be discussed.

Keywords: Cocaine, cocaine contamination, cocaine transfer, cocaine persistence, primary contamination, secondary contamination

1. INTRODUCTION

For several years, Law Enforcement agencies have been conducting research and associated field studies for the purpose of evaluating new and existing narcotic detection instruments. As a result of these successful research projects, several Federal and International law enforcement agencies now have operational programs incorporating the use of different narcotic detection technologies into the everyday routine of their field officers. To date, every agency utilizing these instruments has met with success in the field. As success has been experienced in the field, the natural progression of the programs has led to suspects being indicted on Federal charges of smuggling. The end result is the validity of the technologies being challenged in court. Some of the common arguments used by defense attorneys are: Clients becoming contaminated by touching areas the "real" smuggler touched; and clients becoming innocently contaminated by being in the same area as the smuggler or someone else that was contaminated. The ability to counter these arguments with scientific data needed to be addressed.

In an attempt to determine the validity of the defense challenges as well as to gather information with which to address these arguments, the Narcotic Detection Technology Assessment (NDTA) Team has undertaken a series of field studies. The aim of these tests is to begin to form an understanding of how a person becomes contaminated and how, if at all, this contamination is transferred to surrounding people and objects. The overall goal is to increase the ability of Law Enforcement Agencies (LEAs) to find smuggled narcotics, over a variety of scenarios and conveyances, while maintaining the ability to determine which individuals are involved in the smuggling activity and which are not. The first field study in this series was held in Miami, FL in January 1995. The second field study was held in Miami, FL in March 1995. A summary of these two tests was presented in October 1995 at the ONDCP Conference in Nashua, NH. Based upon questions resulting from these first two studies, a third study in this series was scheduled. This third test was held in Miami, FL beginning on 19 February and ending on 23 February, 1996. To further our base of knowledge, several studies were undertaken, each with its own objective. A listing of these tests and their respective objectives are as follows:

- *Contamination and Transfer Study* - Previous iterations of this test indicate that primary contamination (i.e., objects becoming contaminated as a result of being touched by someone who has handled drugs) does occur. To date, there have been no instances of secondary contamination (i.e., someone becoming contaminated as a result of touching an object that has been primarily contaminated). These previous iterations have involved individuals contaminating themselves at a seizure museum. The objective of this study was to determine whether similar findings would result if the test subjects contaminated themselves with several kilos of actual seized cocaine.

- *Secondary Contamination Study* - The objective of this study was to determine if people can be secondarily contaminated. If so, is there some level below which people do not become contaminated. Based upon findings in previous persistence studies (i.e., cocaine imbedded in pores of skin and being exuded over time), this test was conducted under two conditions, with and without hand washing by the contamination subjects.

- *Contamination Persistence Study* - The objective of this test was to monitor the behavior of cocaine contamination on a person's hands for an extended period of time. Results of previous tests indicate that cocaine becomes imbedded in the pores of the subjects' hands and, over time due to natural physical processes, is exuded out. This study tested several different variations of hand conditions: no washing whatsoever, normal washing, and severe washing (adverse conditions). The goal was to determine, for each condition, whether the contamination levels off to some nominal value or whether it actually decreases to zero.

1.2 Test Team Organization

The test team was comprised of members of the NDTA Team representing the following organizations:

- United States Coast Guard Research and Development Center
- United States Customs Service, Applied Technology Division
- Revenue Canada Enforcement Directorate
- Revenue Canada Laboratory and Scientific Services Directorate
- Office of National Drug Control Policy
- Houston Advanced Research Center
- Argonne National Laboratory

Additionally, observers representing the following organizations were also present:

- Australian Customs Service
- Old Dominion University

2. CONTAMINATION AND TRANSFER STUDY

2.1 Test Objectives

Previous iterations of this test have involved individuals contaminating themselves at the Seizure Museum at the Port of Miami followed by a two hour car ride. Results indicated that cocaine does transfer to surrounding objects during a car ride. The level to which the test subjects were contaminated has never been quantified. For this reason, it is unknown if these results accurately represent a smuggling event. For this study, the test was redesigned to reproduce a real act of smuggling drugs. The objective of this study was to determine whether similar findings would result if the test subjects contaminated themselves with several kilos of actual seized cocaine.

2.2 Methodology

This test replicated the previous two tests in every manner except the contamination step. This step was carried out at the U.S. Customs Seizure Vault. This facility is used to securely store seized narcotics until they are authorized for destruction. The contamination was accomplished by having each of the test subjects carry as many kilo bricks of cocaine as they could (approximately 10 kilos). The test subjects handled the bricks by cradling them in their arms and holding them against their bodies. The subjects carried the kilos to the opposite side of the vault and stacked them on the floor. This process was repeated for 5 minutes. Following is a step by step review of the entire test:

1. Determination of Background Levels. Two "personal" wipes were obtained from each participant prior to starting work. The purpose was to determine individual baseline levels of contamination. These two personal wipes were taken from each persons hands/face and clothes.

2. The car that the test subjects rode in was also wiped prior to the test to determine its baseline level of contamination. The areas most likely to be touched by or come into contact with the test subjects were sampled.

3. The three designated subjects then contaminated themselves with the kilos of cocaine in the manner described above. After contamination, a sample of each subject's hands/face and clothes were taken to determine the level of contamination resulting from the contamination process.

4. The three contamination subjects then drove for approximately 3.5 hours in the presampled car. The car ride commenced approximately 2 hours after the contamination event. During the car ride, the air conditioning was turned on and the windows were rolled up. Additionally, an uncontaminated control subject rode with the contamination subjects to determine if any cross contamination might occur. During the ride, the four occupants changed seats with one another at roughly equal intervals. In this manner, each individual occupied each of the four seats (drivers seat included) at some point in the ride. Each individual and the area he just occupied were sampled after each change prior to sitting down in his "new" seat for the next portion of the ride. Each person and each area was again sampled at the end of the ride. A chase car with dedicated samplers accomplished all the sampling.

2.3 Results

Each of the contamination subjects became contaminated as a result of carrying the kilos of cocaine. The levels of contamination were, to a surprising degree, very consistent. However, this may indicate that they were at, or approaching, the saturation level of the analytical instrument (in this case the Ionscan, manufactured by Barringer Instruments). The control subject did not become contaminated during any part of the test. The test was structured such that the control subject drove the car last (i.e., after every contamination subject had already driven). Even this purposeful manipulation of the seating arrangement did not result in any contamination on the person of the control subject. Figures 1 & 2 highlight the personal contamination, both hands and clothes, throughout the test. Comparing these two data sets reveals an interesting anomaly. As the test progresses, the contamination level on the subjects' hands decreases significantly over time. However, the contamination level on the subjects' clothes only decreases slightly over time.

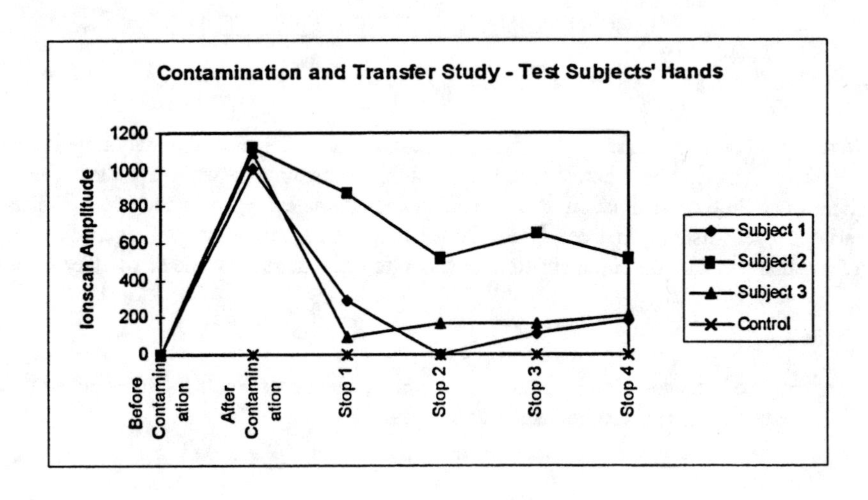

Figure 1 - Contamination and Transfer Study - Subjects' Hands

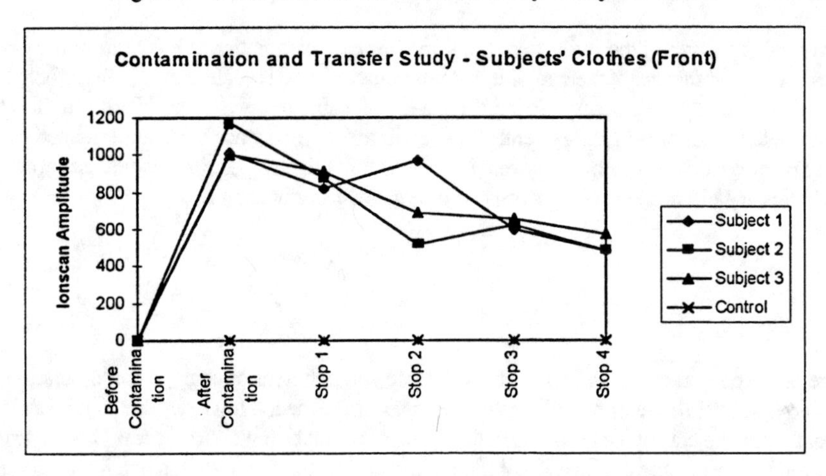

Figure 2 - Contamination and Transfer Study - Subjects' Clothes

In no instances did the backs of the contamination subjects' clothes become contaminated. This result suggests that the cocaine contamination did not migrate (i.e., *transfer*) to other parts of the test subject under this test condition.

This result suggests that the contamination on the subjects hands was of the particulate form. As the test progressed this particulate contamination was either worn off the subjects' hands or absorbed into the skin. Since hands have a relatively smooth surface and small area in comparison to clothes, it is possible that the repetitive hand wiping removed most of the cocaine. However, the fact that neither the negative control subject or the vehicle became contaminated lends weight to the argument that the cocaine on the contamination subjects' hands was actually being absorbed into the skin. If the cocaine was simply being wiped off the hands, then at some point there had to be loose contamination on the surface of the hands. This loose contamination would have been transferred as a result of touching objects in the car. Although the results from other tests suggest that this absorbed cocaine will begin to be exuded over time, these results suggest that this type of contamination does not transfer easily.

This conclusion is further supported by the fact that *the car did not become generally contaminated.* The only area that became contaminated was the right front inside and outside door handles. Contamination was found on this area on two occasions; after the first leg of the drive, amplitude 302; and after the third leg of the drive, amplitude 104. Considering the original level of contamination on the test subjects and the amount of movement that took place inside the vehicle over almost four hours, this is a significant result. It is the beginning of a true understanding of the fate and behavior of cocaine contamination on a person's hands.

3. CONTAMINATION PERSISTENCE STUDY

3.1 Test Objectives

Results from the previous studies in this series indicated that cocaine contamination will persist on a person's hands for a period of time. However, it has not been determined how long this contamination will persist for. The objective of this test was to monitor the behavior of cocaine contamination on a person's hands for an extended period of time. This test was repeated with several different variations, no washing whatsoever, normal washing, and severe washing (adverse conditions). The goal was to determine, for each condition, whether the contamination levels off to some nominal value or whether it actually decreases to zero.

3.2 Methodology

This test monitored the behavior of cocaine contamination on a person's hands for an extended period of time under three different conditions. These three different conditions are as follows:

- *No Washing*
- *Normal Washing*
- *Adverse Conditions* (Showers & Hot Tub)

The three test subjects for this portion of the study were the same three contamination subjects from the Contamination and Transfer Study. This study was simply commenced at the end of the Contamination and Transfer study. Therefore, the beginning of this study is considered to be the point at which the subjects carried the kilos of cocaine in the Seizure Vault. In this manner, this portion of the study is also a true representation of a smuggling event. Several hours of persistence were experienced (the time the Contamination and Transfer study was being conducted) before any washing was accomplished. During the test, separate wipe samples were taken from each subject's hands and clothes at roughly two hour intervals during waking hours.

3.3 Results

3.3.1 No Hand Washing

The purpose of this portion of the test was to study the fate and behavior of cocaine contamination on a person's hands over an extended period of time when absolutely no hand washing was accomplished. However, due to normal circumstances in daily life, at least one instance of the test subject wiping his hands with a moistened paper towel occurred. This occurred before the sample was taken at the 2.75 hour mark. The marked decrease in contamination can be attributed to this hand wiping. In a sense, this persistence study can be considered one conducted during a period of ordinary daily life. However, to differentiate this particular test from the other persistence studies, it will still be referred to as the "no hand washing iteration". The results of this iteration of the test are displayed in Figure 3.

The results from this portion of the test further support the hypothesis that cocaine is absorbed into the skin and later exuded out. At the 3.75 hour mark, the contamination level had, unexpectedly, decreased to an undetectable level. However, over the next six hours this contamination level rises to a maximum of 205. This same pattern is seen again at the 20 hour mark. A note must be made about this data point, though. This sample was taken approximately 1.5 hours after the test subject woke up and got dressed in the same clothes worn the previous day. This level may be due to contamination coming back off his clothes onto his hands. A sample of his clothes taken at the same time had an IMS amplitude of 412, further supporting this theory.

3.3.2 Normal Hand Washing

The purpose of this portion of the test was to study the fate and behavior of cocaine contamination on a person's hands over an extended period of time when normal washing and showering habits are observed. This study introduces the variable of normal washing. In all instances of washing, drying was accomplished with a towel, not air drying. The results of this iteration of the test are displayed in Figure 4.

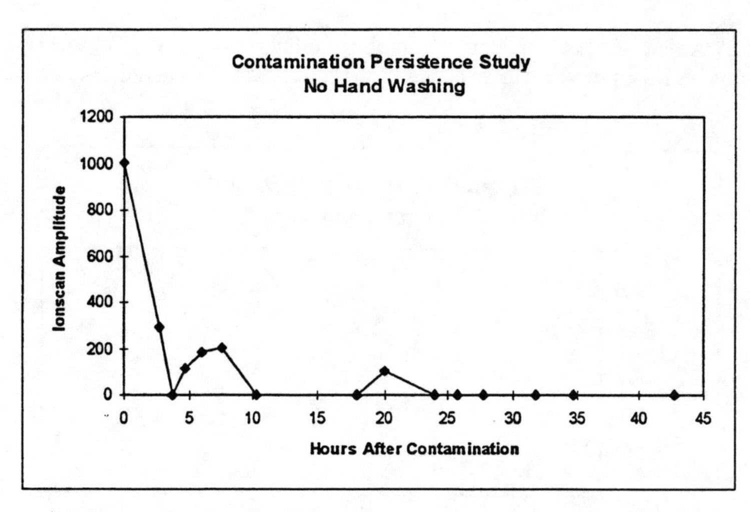

Figure 3 - Contamination Persistence Study - No Hand Washing

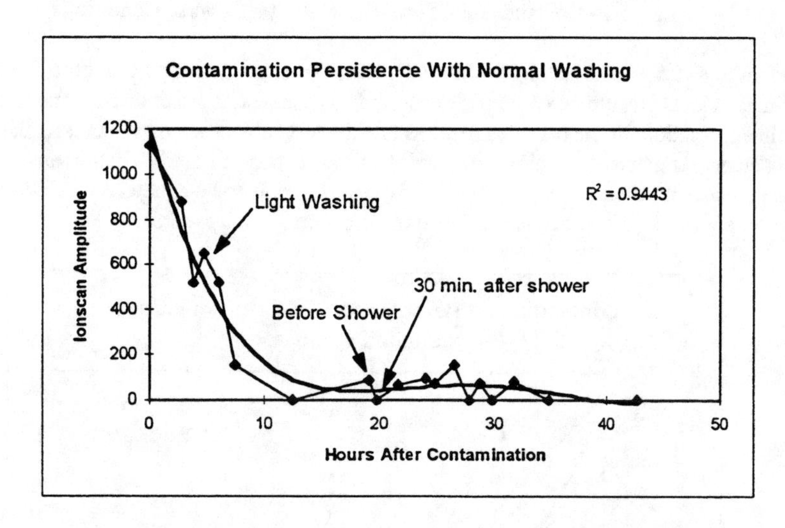

Figure 4 - Contamination Persistence Study - Normal Washing

Once again, the results of this test support the hypothesis that cocaine is absorbed into the skin and later exuded out. During the course of this test the test subject had two washing events that had significant results. These washings, a hand washing and a shower, occurred at the 7 hour mark and at the 19 hour mark, respectively. These events are highlighted on Figure 4. As expected, the contamination level on the test subjects hands decreased markedly as a result of the washing. In both cases, the contamination level went to zero (or below the preset detection threshold on the IMS) for a short period of time. After the temporary decrease, however, the contamination increased again. This series of events can be explained in the following way. The washing washed off any contamination that was on the surface of the skin. As a result, there was no surface contamination for a short period of time. As the cocaine began to be exuded from the pores of the skin, however, the contamination level again increased. The tail end of the curve in Figure 7 seems to indicate that the overall up and down fluctuations in the contamination level decrease over time. The conclusion is that the contamination ultimately decreases to zero.

3.3.3 Adverse Conditions

The purpose of this portion of the test was to study the fate and behavior of cocaine contamination on a person's hands over an extended period of time when subjected to severe washing conditions. These severe conditions are defined as a shower and a 20 minute bath. The impetus behind this test condition is a result from the previous study in this series. In that test, the cocaine contamination on a test subject's hands survived these conditions. The results of this iteration of the test are displayed in Figure 5.

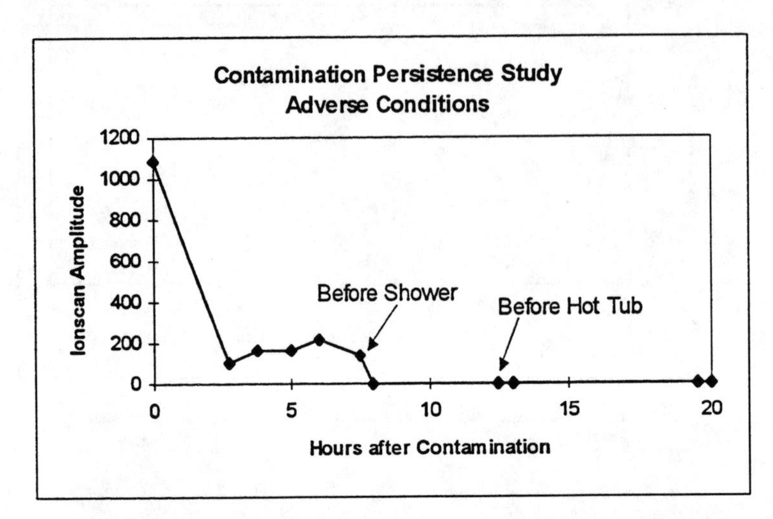

Figure 5 - Contamination Persistence Study - Adverse Conditions

As can be seen in Figure 5, the test subject exhibited a normal contamination pattern until the adverse conditions washing took place. The contamination level went to zero once this occurred. These results are in direct contradiction to the results from the previous test. A possible explanation for the previous result is that the test subject wore the same clothes the following day after all the washing, thereby recontaminating himself. A closer inspection of the portion of Figure 5 before any washing occurred once again supports the hypothesis that cocaine is absorbed into the pores of the skin only to be exuded out. Figure 6 is a graph of this portion of Figure 5 out to the 7.5 hour mark. Note the classic contamination persistence pattern.

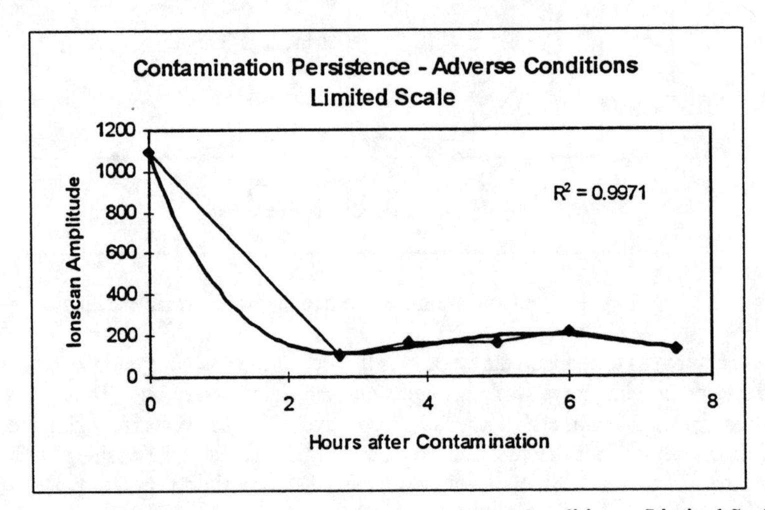

Figure 6 - Contamination Persistence Study - Adverse Conditions - Limited Scale

A general note must be made about the entire persistence study. As can be seen from the results of every iteration of the study, the 0 to 2.75 hour mark is critical. Two of the three test subjects experienced large decreases in contamination over this period. A further review of the notes indicate that a lunch break was taken during this period, from the 1 to 2 hour mark. Lunch was taken at a restaurant that served sandwiches. Obviously, some hand wiping occurred during this period. It is unknown what the effects of this hand wiping were, but is a safe bet to assume that some of the recorded decrease in contamination can be attributed to this. However, eating is a normal function in daily life. Therefore, this event in no way invalidates the test. It is reported, though, in keeping with good scientific practices.

4. SECONDARY CONTAMINATION STUDY

4.1 Test Objectives

Results from the previous studies in this series indicated that, in no instances, was there an occurrence of secondary contamination, only primary contamination. Primary contamination is defined as an area becoming contaminated as a result of being touched by a person who had direct contact with narcotics. Secondary contamination is defined as a person becoming contaminated as a result of touching an area with primary contamination.

The objective of this study was to determine if people can be secondarily contaminated. If so, is there some level below which people do not become contaminated. Based upon findings in previous persistence studies (i.e., cocaine imbedded in pores of skin and being exuded over time), this test was conducted under two conditions, with and without hand washing by the contamination subjects.

4.2 Methodology

The thrust of this test was to primarily contaminate several handheld items of different composition. Negative control subjects in turn handled each one of these objects to determine if they could become secondarily contaminated. The following test procedure was used:

1. Determination of Background Levels. One personal wipe was obtained from each participant's hands prior to the start of the test to determine baseline levels of contamination.

2. Three test subjects then contaminated themselves on several items in the Seizure Museum at the Port of Miami. Each test subject's hands were then wipe sampled to determine original levels of contamination for the test.

3. Each test subject then handled several objects, in turn (i.e., primary contamination). Each subject was "assigned" an identical set of objects. The particular objects were as follows:

   - 9" x 9" Aluminum Plate (1/4" thick)
   - Suitcase Handle (Leather or Hard Plastic)
   - 6" x 6" Piece of Denim Cloth
   - 4" x 6" Piece of ½" Plywood
   - Telephone Handset

   After handling all of the objects, the test subjects' hands were wipe sampled again to determine the remaining contamination level.

4. Three negative control subjects were then assigned one set of contaminated objects each from step 3. The negative control subject then touched one of the contaminated objects at a time. After touching each object, the subject's hands were wipe sampled and analyzed to determine the level of contamination (i.e., secondary contamination), if any, as a result of touching the object. If the subject's hands were contaminated, his hands were washed with soap several times and verified to be contamination-free before touching the next object. This process was repeated for each object in the set and for each of the three sets of objects.

5. Each object was then wipe sampled to determine the level of primary contamination.

On consecutive days, two iterations of this test were accomplished. Following is a description of each test condition:

- *No Hand Washing* - The test subjects contaminated themselves on items in the Seizure Museum. Objects were sampled directly thereafter.

- *With Hand Washing* - The test subjects contaminated themselves on items in the Seizure Museum. After 20 minutes, the test subjects rinsed their hands with water only. In this manner, all "loose" contamination was washed off the surface of their hands. The objects were touched 75 minutes after hand washing.

4.3 Results

4.3.1 No Hand Washing

The results from this test counter the results from previous tests in that secondary contamination did occur. The results begin to define the limit for secondary contamination. In some cases, the secondary contamination level was very high. Table 1 summarizes the results of this test. Table 1 is formatted such that, reading from left to right, one is able to track the contamination on the test subjects' hands to the resultant contamination on each object (primary contamination) to the contamination that came off of each object onto the negative control subjects' hands (secondary contamination). The table ends in the rightmost column with the final contamination level on each test subjects' hands (average of left and right hands) after touching all of the objects.

Test Subject	Primary Subjects Hands After Contamination	Object	Objects After Being Touched by Primary Subjects	Secondary Subjects Hands After Touching Objects	Average Primary Subjects Hands After Touching Objects
1A	1300 *	Plate	1059	918	1059
		Handle	425	0	
		Cloth	1006	310	
		Wood	885	339	
		Phone	995	510	
1B	1300 *	Plate	1039	1300 *	1116
		Handle	963	969	
		Cloth	1300 *	963	
		Wood	974	1082	
		Phone	1300 *	976	
1C	1300 *	Plate	903	903	1098
		Handle	860	747	
		Cloth	975	521	
		Wood	946	599	
		Phone	815	276	

Table 1 - Summary of Secondary Contamination Study - No Hand Washing

* Note: This indicates an Ionscan 250 Saturation condition. Based upon the instrument calibration curve, 1300 was assigned as the appropriate saturation level value.

As can be seen in Table 1, secondary contamination occurred in all cases except for one. The one instance in which secondary contamination did not occur happened to be the sample in which the lowest level of primary contamination was found.

4.3.2 With Hand Washing

The addition of one extra step in the test procedure, rinsing hands with water, resulted in totally different results. In no instance was secondary contamination found. Table 2 summarizes the results of this test.

Test Subject	Primary Subjects Hands After Contamination *	Object	Objects After Being Touched by Primary Subjects	Secondary Subjects Hands After Touching Objects	Average Primary Subjects Hands After Touching Objects
2A	301	Plate	190	0	327
		Handle	0	0	
		Cloth	71	0	
		Wood	201	0	
		Phone	0	0	
2B	450	Plate	175	0	410
		Handle	0	0	
		Cloth	0	0	
		Wood	83	0	
		Phone	0	0	
2C	0	Plate	0	0	72
		Handle	0	0	
		Cloth	0	0	
		Wood	0	0	
		Phone	0	0	

Table 2 - Summary of Secondary Contamination Study - With Hand Washing

* Note: This contamination level represents the level on the primary subjects hands after contamination, hand wash and air dry 20 minutes later, followed by a 75 minute wait prior to handling the objects.

4.3.3 Overall Results

Contamination-wise, the only difference between the two iterations of this test was how the contamination was deposited on the surface of the test subjects' hands. In the no hand washing iteration, the contamination consisted of particulates on the surface of the hands. The hand washing iteration effectively did away with the surface contamination. The only contamination left was of the type that had formed a thin film on the skin and/or had been absorbed into the subjects' hands and had begun to be exuded out. The results indicate that this type of contamination is much more difficult to transfer. One of the goals of this portion of the study was to determine if it is possible to highlight a contamination level below which secondary contamination will not occur. In order to make this determination, the results of both iterations of this test were graphed. This graph is presented as Figure 7. For every sample taken in this study there are three points on this graph: The contamination on the test subjects hands prior to touching the objects; the primary contamination level on the objects after being touched by the test subject; and the resulting secondary contamination level on the secondary test subjects' hands after touching the contaminated objects.

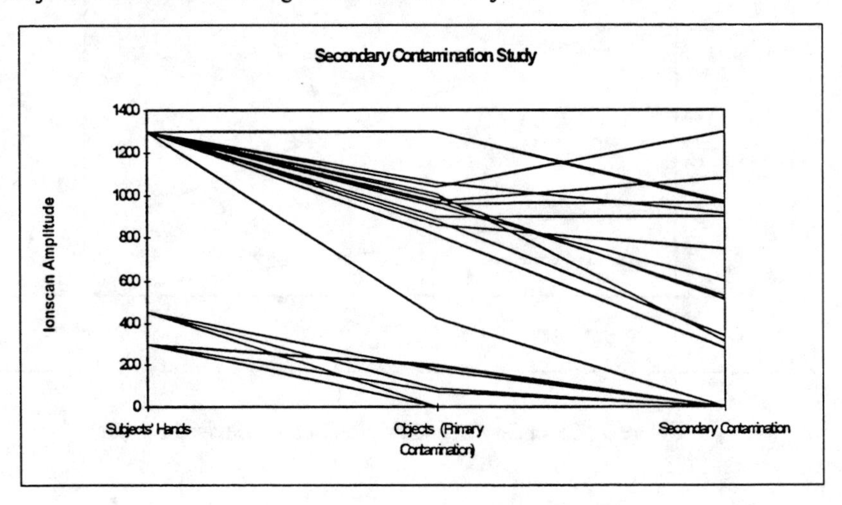

Figure 7 - Secondary Contamination Study

Figure 7 is obviously a very busy figure. A closer inspection reveals that there is no "overlap" between samples in which secondary contamination occurred and samples in which secondary contamination did not occur. As a result, Figure 7 may be segregated into three areas: An area containing samples in which secondary contamination occurred; An area containing samples in which secondary contamination did not occur; and an area in which we have no data. This segregation results in Figure 8.

An inspection of Figure 8 reveals that all samples that had a level of primary contamination on the objects less than an amplitude of 425 did not contribute to any secondary contamination on the control subjects' hands. Conversely, all samples that had a level of primary contamination on the objects greater than an amplitude of 815 resulted in secondary contamination on the control subjects' hands. We have no data on the area between primary contamination levels of 425 and 815. An inspection of Figure 7 reveals that a single data series results in the upper boundary of the "no secondary contamination" area. This data series has a far greater amplitude than any other data series in this set. Therefore, if this data series is considered an "outlier", Figure 8 may be revised. This revised summary results in Figure 9. This figure may more accurately reflect the areas that may or may not result in secondary contamination. Further research may lessen the gap where we have no data and possibly better define areas of overlap.

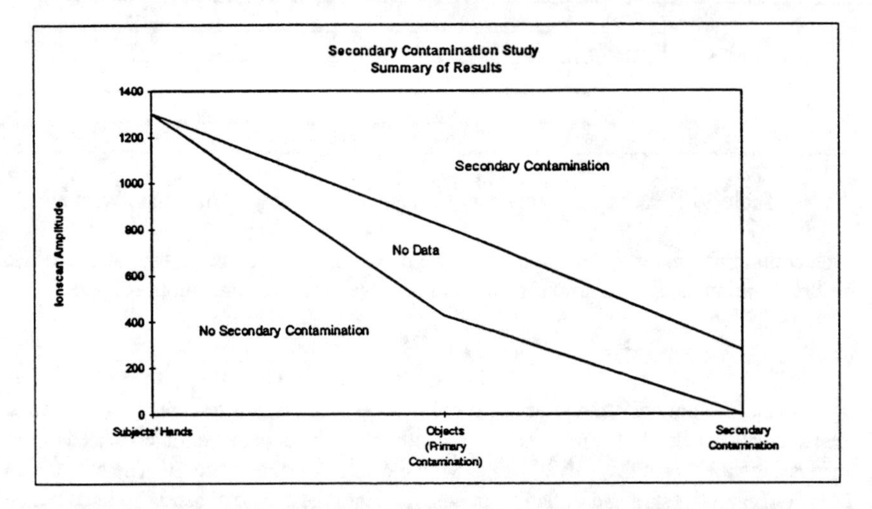

Figure 8 - Secondary Contamination Study - Summary of Results

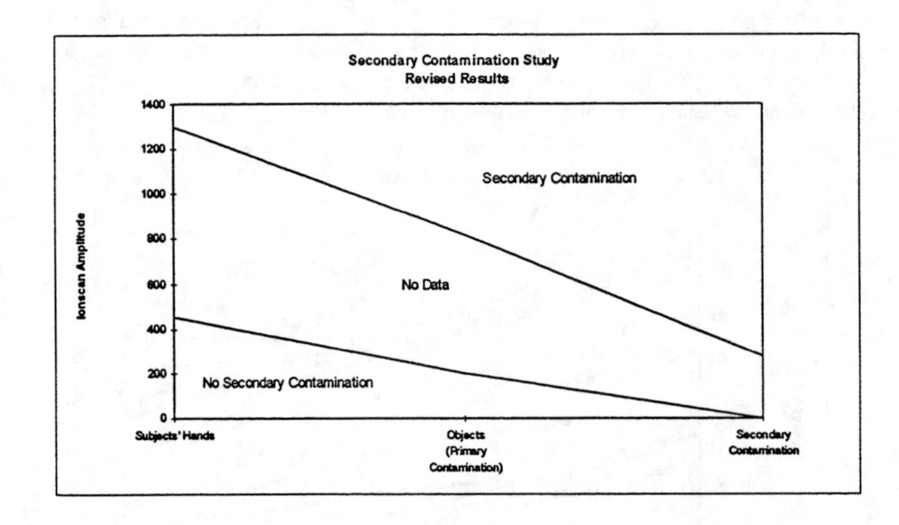

Figure 9 - Secondary Contamination Study - Revised Results

5. SUMMARY AND DISCUSSION

The results of this field test have both helped to solidify our perceptions on certain topics as well as to provide additional information which has helped to refine some of our perceptions. Following is a test-by-test summary of the entire field study:

- Contamination and Transfer Study: Even though all three test subjects became highly contaminated, there appears to be minimal transfer to surrounding people and objects. The overriding conclusion is that, in the context of this study, *a person who becomes contaminated as a result of carrying bulk narcotics in the smuggling process does not easily contaminate areas that he touches.* This is not to say that primary contamination is not possible. It does say, however, that *there is enough scientific data to counter any argument that states that primary contamination is widespread on vessels involved in smuggling.* This statement supports U.S. Coast Guard operational data which shows that, even on vessels involved in smuggling large amounts of cocaine, contamination is not widespread in general areas.

- Secondary Contamination Study: The results from this better define contamination levels required for secondary contamination to occur. Combining the results from both portions of the test results in a plot containing no overlap between samples that resulted in secondary contamination and samples that did not contain secondary contamination. Although the data indicates that there is a definite demarcation between primary contamination levels that do and don't contribute to secondary contamination, logic mandates that there is some "grey area" in which there is an overlap of results. The end goal of this test is to determine this overlap area. It does seem possible that we may ultimately be able to definitively state primary levels above which secondary contamination always occurs and primary levels below which secondary contamination never occurs.

- Contamination Persistence Study: The results from all three portions of this test indicate that cocaine contamination may be absorbed into the pores of the skin. Over time, due to natural physical processes, this absorbed contamination is exuded back out. This conclusion is based upon the contamination persistence curves. In all instances where the contamination survived for an extended period of time, there was the trend in which a high level of contamination decreased markedly only to increase again over time. The most logical explanation for this increase in contamination is from the pores of the subjects' hands. This test was a realistic re-creation of an actual smuggling event. Based upon the persistence curves from each condition in this study, it is a valid conclusion to state that the cocaine contamination level on a person's hands rapidly levels off to a nominal value. Over time, this contamination level will eventually decrease to zero. Dependent upon the amount of washing that the subject accomplished, the time required to reach zero may be as short as several hours or as long as several days.

6. ACKNOWLEDGEMENTS

The Narcotic Detection Technology Assessment Team would like to thank the following individuals for their assistance. Without their help and logistical support, several portions of this field study would not have been possible.

- Mr. Dave McKinney - U.S. Customs Contraband Enforcement Team Director, Port of Miami.

- Mr. Jayson Ahern - Port Director, U.S. Customs Service, Miami International Airport.

- Mr. Jeffry Baldwin - Port Director, U.S. Customs Service, Port of Miami.

- U.S. Coast Guard Base Miami Beach.

Traces of illegal drugs on body surfaces - indicator for consumption or dealing?

Franz Aberl, Johannes Bonenberger, Ralf-Peter Berg, Rudolf Zimmermann
Securetec GmbH, Rosenheimer Landstr. 129, D-85521 Munich/Ottobrunn,
Tel.: +49/89/607-23103, Fax: +49/89-607-29182

Hans Sachs
Institute for Legal Medicine, Ludwig-Maximilians-Universität München, Frauenlobstr. 7a, D-80337 Munich

ABSTRACT

Customs investigation and drug enforcement services are interested in a rapid and reliable identification of smugglers (e.g. body packers) and dealers. In contrast workplace testing and traffic controls are aiming at the detection of intoxicated persons via the determination of illegal narcotics in body fluids like urine or blood.

DRUGWIPE is a pen size, test strip based immunochemical detector for narcotic contaminations on surfaces. It is extremely simple to apply and takes about two minutes to read test results without depending upon any further technical means.

This paper describes the applicability of *DRUGWIPE* to identify drug smugglers or dealers as well as consumers. With respect to the situation and the initial suspicion the test indicates handling as well as consumption. In cooperation with the Institute for Legal Medicine in Munich suspicious drivers were examined with *DRUGWIPE* for the abuse of illegal narcotics. Test results from this test series are presented and compared with the results from the blood or urine analysis. The question wether the detected traces of illegal narcotics on the body surface of suspicious drivers are coming from transpiration or external contamination are discussed.

Keywords: Illegal drugs, customs service, traffic control, consumption, body packing, surface contamination, immunosensor, sweat

1. INTRODUCTION

Officers involved in the fight against the abuse of illegal drugs often have an intuitive feeling or even some unspecific indications that something is "wrong" with a person under examination. Their problem is to generate from first, unspecific indications a reasonable cause which allows the officer to proceed to more strict methods of investigation.

In such situations information about the presence of traces of illegal drugs on a certain surface can help to confirm or contradict an initial suspicion. Residues of illegal drugs on body surfaces can provide different kinds of information. Smugglers or dealers usually contaminate themselves on their hands or fingers on a very low level with an illegal narcotic. Therefore such traces can indicate that someone is possessing drugs-of-abuse. Such traces can also be helpful to correlate seized narcotics to a person under suspicion.

Furthermore surface contaminations on hands or fingers can be exploited to indicate body packing. If the drug packets - prepared for swallowing or stuffing - are sufficiently contaminated a body packer will suffer from contamination on his fingers or - in the case the drug packets are feeded by another person he will take up, metabolize and excrete drugs comparable to a consumer. This gives the officers the possiblity to check body fluids like urine, blood, sweat or saliva for the presence of illegal drugs.[1]

Two different operational conceptions both correlated with the detection of illegal drugs on body surfaces were evaluated. The German Customs Service proved the applicability of *DRUGWIPE* to detect body packers based on invisible residues on their hands or fingers. Whereas a German Institute for Legal Medice is evaluating *DRUGWIPE* concerning it´s applicability to identify consumers of illegal drugs under traffic control conditions. This study is based upon the measurement of drug residues excreted in sweat.

2. CONSTRUCTION AND PERFORMANCE OF *DRUGWIPE*

DRUGWIPE is a universal detector for narcotic contaminations on surfaces. Four main groups of illegal drugs are to be detected:

- Cocaine
- Opiates
- Cannabinoides
- Amphetamines (Amphetamine, Methamphetamine, Methylendioxymethamphetamine (MDA), MDMA, MDE)

The *DRUGWIPE* field tests are characterized by their rapid response times and a sensitivity in the low nanogram range. Technical data are summarized as follows.

Characteristics	Value	Unit	Remark
• Sensitivity	5 - 50	ng	depending on the detectable drug
• Analysis time	2	min	includes sampling and measuring
• Size	13 x 1 x 2	cm^3	
• Weight	12	g	

Table 1: Main technical characteristics of the *DRUGWIPE*

DRUGWIPE's most important feature is it's high sensitivity. The lower limit of detection gives the drug enforcement officer on-site the opportunity to quickly detect invisible traces of illegal drugs. Traces of illegal narcotics can even be detected after someone has washed his hands.

Figure 1 illustrates the construction principle of the drug detection kit *DRUGWIPE*. *DRUGWIPE* consists of 3 main parts, each part fulfilling a different purpose:

- The wiping section is a tool for sampling drug particles from different kinds of surfaces.

- The detection element contains a test strip based immunosensor.[❶]
- The tap water container also acts as protection cap.

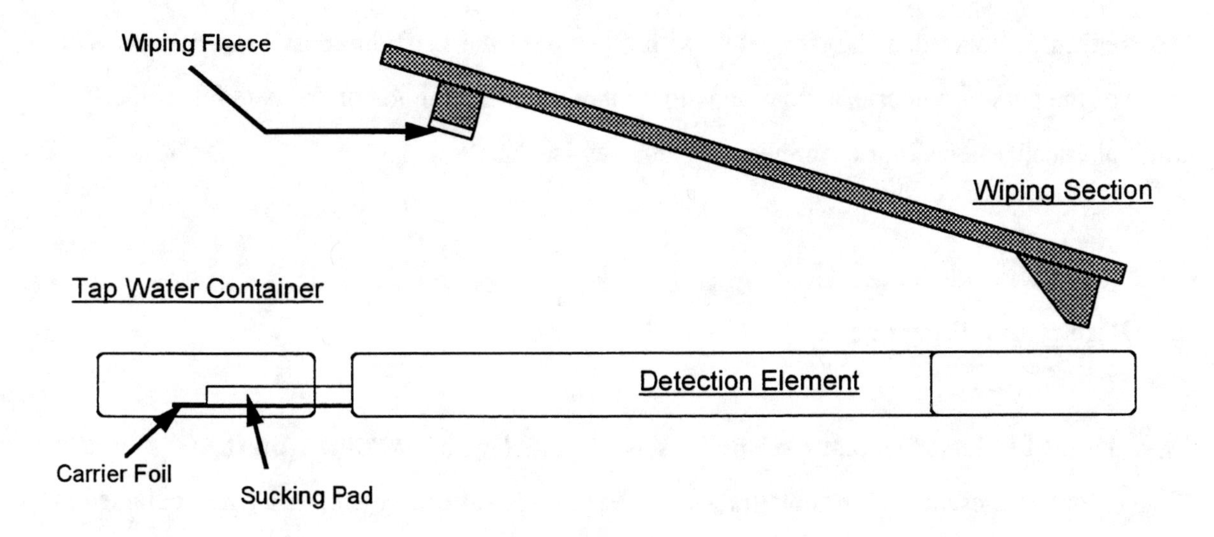

Figure 1: The construction principle of the *DRUGWIPE* test system

The test strip itself consists of different functional elements containing the immunochemical reagents necessary for the detection and identification of the different drugs. The basic working principle and the application steps of *DRUGWIPE* are described elsewhere.[2, 3]

3. DETECTION OF BODY PACKERS

3.1 Organisational and methodical details of the Customs´ field test series

The German Customs Service evaluated *DRUGWIPE* concerning it´s usefulness for the detection of body packers and smugglers. The drug enforcement groups at the airports in Frankfurt,

---

❶ The technological basis of *DRUGWIPE* is the test strip Frontline® which is already aproved by the FDA for urine testing.

Munich and Cologne were equiped with *DRUGWIPE*s and trained in the application of the test system. The persons tested were preselected with respect to the experience of the Customs officers.

To confirm or contradict the test result with *DRUGWIPE* the Customs Service either analyzed urine of the persons under examination with immunological methods or the persons themselves were physicaly searched or examined with an x-ray machine.

3.2. Results and discussion

In a period of about 3 months 65 persons were tested with *DRUGWIPE* on their hands and fingers for the presence of invisible traces of illegal drugs. a positive test result was obtained in 11 cases (17%). Table 2 summarizes the results of these 11 positive cases together with the results of the different methods of confirmation which were applied.

8 out of the 11 positive persons (73%) were identified as body packers. The urine of two body packers was negative. These persons had either no or only an insufficient quantitiy of drugs or it´s metabolites in the urine.

In one case a suspicious person was identified by a positive urine test result as a consumer of illegal narcotics. In only two cases the *DRUGWIPE* test result could not be confirmed by a confirmatory methods. In these cases the measured drug residues are assumed to originate from unintentional contamination, e.g. by touching contaminated subjects or these persons consumed drugs several days previously and were still contaminated.

In all other cases (54 out of 65, 83%) the test result with *DRUGWIPE* was negative. These results were confirmed with at least one of the methods mentioned above.

No	*DRUGWIPE* "Cocaine"	Urine test	X-ray	Seizure	Remark
1	+	-	+	yes	body packer
2	+	+	-	yes	body packer
3	+	-	-	no	contaminated
4	+	-	-	no	contaminated
5	+	-	+	yes	body packer
6	+		+	yes	body packer
7	+	+	+	yes	body packer
8	+	+	+	yes	body packer
9	+	+	+	yes	body packer
10	+	+	-	no	consumer
11	+	+	+	yes	body packer
12-65	-	-	-	no	-

Table 2: A comparison between the results of the *DRUGWIPE* "Cocaine" test system and different confirmatory methods. The results originate from a field test campaign of the German Customs Service.

3.3 Conclusions

DRUGWIPE fulfils several important demands to a test system for illegal drugs fully usuable in the field. In the context of personal examinations several important aspects can be conluded from the test series conducted so far. Body packing or smuggling can be indicated quick and reliable. This helps the officer to generate out from his first, unspecific intuition a reasonable cause for further, more strict methods of investigation. A second, and at least as important aspect is that an initial suspect can be reliably contradicted in the field. This avoids the application of further methods of investigation which may require expensive equipment, a personal consent, or an intrusion into one´s privacy.

4. THE IDENTIFICATION OF INTOXICATED DRIVERS

4.1 Organisational and methodical details of the test campaign

Actually, in Germany an adoption of the street traffic law is under preparation. This adoption regulates that driving under the influence of illegal narcotics is punishable because it violates valid regulations. For the traffic police this adoption of the new regulations leads to the problem how to recognize and identify drivers under the influence of illegal drugs in traffic control situations. The traffic police in Germany (especially in Bavaria) is currently trained in recognizing drug influenced drivers by means of particular mental and physical criteria like extreme moods or abormal local or temporal orientation. Actually no roadside test is available on the market which can detect the abuse of narcotics in a non-invasive manner rapidly, reliably and without the consent of a suspicious driver.

Historically the most reliable and simple way to test for the abuse of drugs with the consent of a person is to measure illegal drugs or its´ metabolites in urine. This test series described is is aimed at the detection of drug residues in sweat. The amounts excreted via sweat are rather low and have been, up to now, only measurable by laboratory based methods. However sweat can be sampled in more non-invasive manner and is therefore, with the advent of immunochemical detection technologies, a more suitable matrix for testing in the field.

At the Institute for Legal Medicine in Munich blood or urine of drivers who are potential abusers of drugs are screened for the presence of legal or illegal drugs. Out of this population drivers on a voluntarily basis are examined with *DRUGWIPE* for the presence of detectable residues of illegal drugs in their sweat. The sweat was sampled in the armpits to guarantee that no cross contamination is measured. If necessary persons are tested with two or three types of the *DRUGWIPE*, but in most cases only one type of the *DRUGWIPE* field test system (either Cocaine, Cannabis or Opiates) is applied.

To confirm the analytical results of the *DRUGWIPE* field test system laboratory based instrumental methods of analysis where applied. For the examination of urine or blood a two step procedure was used. First a rapid screening of the human body fluids was performed based on immunochemical methods (Fluoreszenz Polarization Immunoassay, FPIA). Combinations of Gaschromatography and Mass Spectroscopy (GC/MS) or High Performance Liquid Chromatography and Diode Array Detection (HPLC/DAD) acted as confirmatory methods.

4.2 Results and Discussion

The following results in tables 3 to 5 originate from the examination of a total group of 34 voluteers. Each table compares the test results with the *DRUGWIPE* system with the test result of the blood or urine analysis.

The highest number of results up to now is available from the *DRUGWIPE* "Opiates" test. 27 persons where investigated with *DRUGWIPE* "Opiates". In 19 cases a positive test result was obtained. 8 persons were negative.

OPIATES		Confirmatory method	
		+	-
DRUGWIPE	+	**17**	**2**
	-	**5**	**3**

Table 2: Comparison of the test results with the *DRUGWIPE* "Opiates" test and the result from the blood or urine analysis with an instrumental analysis method for confirmation.

In 20 out of 27 cases (74%) the results indicated by *DRUGWIPE* "Opiates" are matching with the results obtained by the confirmatory methods. In 5 cases *DRUGWIPE* showed a negative result. This can be explained by an insufficient degree of sweating or excretion of illegal narcotics. With respect to *DRUGWIPE* this means, that the sensitivity of the test system is not sufficient to detect such low amounts of drugs in sweat, e.g. present during the very early period after consumption. In two cases a positive *DRUGWIPE* result could not be confirmed by the subsequent laboratory analysis. These people are assumed to be consumers who have still traces of illegal drugs excreted via sweat in their armpitch e.g. because of insufficient washing.

With the *DRUGWIPE* field test system for cocaine and cannabis 16 measurements were performed up to now. With *DRUGWIPE* "Cocaine" a total of 10 tests and with the Cannabis test system a total of 6 tests were conducted. Only in one case the test result of *DRUGWIPE* does not match with the result of the confirmatory method. This case can be explained by a former consumption where drug residues are still present on the skin in the armpits.

These test results document the actual status of the test series in the Institute for Legal Medicine. In summary 34 persons were tested with 43 *DRUGWIPE*s. In 81 % of all cases (35 cases) the *DRUGWIPE* result measured with sweat agrees with the result from the analysis of blood or urine, based on a laboratory method for confirmation.

COCAINE		Confirmatory method	
		+	-
DRUGWIPE	+	**2**	**1**
	-	**0**	**7**

Table 3: Comparison of the test results with the *DRUGWIPE* "Cocaine" test and the result from the blood or urine analysis with an instrumental analysis method for confirmation.

CANNABIS		Confirmatory method	
		+	-
DRUGWIPE	+	**4**	**0**
	-	**0**	**2**

Table 4: Comparison of the test results with the *DRUGWIPE* "Opiates" test and the result from the blood or urine analysis with an instrumental analysis method for confirmation.

Additional to this scientific evaluation of the *DRUGWIPE* test systems, the traffic police in Germany itself has established a field test campaign to test *DRUGWIPE* under roadside conditions. Traffic policemen were trained and equiped with *DRUGWIPE* tests to enable them to apply *DRUGWIPE* under traffic control conditions. Up to now 7 cases are registered where *DRUGWIPE* was used to indicate the policeman on the road wether a person is driving under the influence of illegal drugs or not.

5. SUMMARY AND OUTLOOK

This paper describes two different test series where the *DRUGWIPE* tests are applied to test persons for the abuse of illegal drugs. In two completely different operational environments *DRUGWIPE* was applied to detect residues of illegal drugs on the human skin.

Customs Service as well as the Traffic Police need a rapid and reliable tool to confirm or contradict first, unspecific indications. *DRUGWIPE* can help both enforcement groups to generate out of an initial suspect a reasonable cause which allows them to apply more strict methods of investigation. Both groups can utilize information gained by simply testing the human body surface for the presence of residues of illegal narcotics.

The German Customs Service is subdivided into about 160 different operational units and boarder crossing points. Due to the successful application of *DRUGWIPE* during the test period the German Customs Service introduced *DRUGWIPE* meanwhile country-wide. The *DRUGWIPE* application profile fits ideally to the Customs operational demands to a rapid and reliable tool for decentralized testing.

The roadside test series with the German traffic police as well as the test series in the Institute for Legal Medicine in Munich will be continued until June 1997 or at least until a statistical sufficient number of test results is available. At the beginning of next year the *DRUGWIPE* field test for Amphetamines will be included into these evaluation campaigns.

Important for a successful application of *DRUGWIPE* under different conditions is to combine the specific operational knowledge of the enforcement groups with the test specific application advantages. Only this combination can generate a new powerful tool in the fight against the abuse of illegal drugs.

1. T. A. Gough. "The rapid screening of persons for internally concealed drugs," Proceedings of the International Symposium Drug Detection, Bundeskriminalamt, Volume 4, December 1990, Wiesbaden, Germany, pp. 61-71.

2. R. Zimmermann, J. Bonenberger, R. Hilpert, F. Binder, C. Klein, H. Droste, A. Goerlach-Graw, F. Scheller, A. Makower. "Biosensor technology for the detection of illegal drugs," Proceedings of the symposium "Counterdrug Law Enforcement: Applied Technology for Improved Operational Effectiveness International Technology Symposium", Nashua, New Hampshire, October 1995, pp. 12-35 - 12-39.

3. R. Hilpert, C. Bauer, F. Binder, M. Grol, K. Hallermayer, H.-P. Josel, C. Klein, J. Maier, A. Makower, H. Oberpriller, J. Ritter. Biosensor technology for the detection of illegal drugs (II) - Antibody development and detection techniques. Proceedings of the symposium "Cargo Inspection Technologies" ed. by A. H. Lawrence, July 1994, San Diego, California, pp.128-138.

A Study to Investigate the Trace Levels of Contamination on Surfaces when Narcotic Contraband is Concealed in a Vehicle

Rod Wilson (Graseby Dynamics Ltd)*, Alan Brittain (Graseby Dynamics Ltd).

Abstract.

When a vehicle is used to transport narcotic contraband material trace levels of that material can be found on surfaces of the vehicle, people associated with the vehicle and surfaces they contact. The detection of these trace levels can help to target vehicles associated with the smuggling of the contraband. A study to determine the typical levels of narcotic material that can be detected from these surfaces has been performed by personnel from Graseby, using a variety of drug materials. The size and packaging of the drug materials has been prepared to try to reflects that typically found in smuggling operations. These test show that for all hard drugs (not cannabis) easily detectable traces of drug material (>10ng) can be found on the vehicle, the proxy (vehicle driver) and secondary surfaces handled by the proxy. For detection of cannabis, the condition of the original material had a great bearing on the reliability of detection.

Introduction

Great developments have been made recently in the application of trace detector technology to the problem of finding concealed contraband materials. The two main classification of materials targeted in this way have been explosives and narcotics. It is the aim of such methods of detection to identify objects or people associated with the smuggling of the contraband material by the detection of trace levels of the material on surfaces. A typical scenario might be that where one person prepares a drug concealment, most often wrapping it and handling the drug packages, then conceals the package in an object in which it can be inconspicuously transported . Then the object might be transferred to a second person, the proxy, who transports the object across a border, the most likely search point. It is most often the case that the surfaces which can be searched, baggage, travel documents, vehicles etc., have not been in primary contact with the drug. Indeed, often the search surface has only been contacted by the proxy, who has had no direct contact with the drug package itself and therefore the traces would be present by secondary transfer processes only. It is the aim of the searcher to retrieve the traces of material and the aim of the detection system to identify them. The probability for detection will therefore depend on many parameters,

- The uptake of the contraband material onto the hands of the drug packer.
- The transfer of material onto the primary contact surfaces.
- The uptake of material onto the hands of the proxy from the primary contact surfaces.
- The transfer of material from the hands of the proxy to the secondary contact surfaces.
- The stability and persistence of the traces on each of the surfaces.

- The facility of the sampling medium to capture the trace and then release it into the detection system
- The facility of the detector to capture and identify a response from the trace in the presence of background material.

In this study it was the aim to examine the probability for detection of traces of the drugand study some of the parameters which affect a successful detection, using one particular scenario and a variety of different drug materials. A reasonable scenario was envisaged to be:

1. The vehicle in which the concealment is going to be made is driven into a room where the package is prepared.
2. The drug materials are brought into the room and the drug packer handles the drug materials to pack them as if preparing a concealment. For cannabis this was done ungloved, however for the hard drugs, rubber domestic-type gloves were worn. This was done primarily for reasons of health and safety, however it is not thought to be untypical of the practice of a real drug smuggler. The drug packages made to try to reflect those found in real drug concealments. The composition is also important as often the smuggled drug are not pure, the active compound often only 10 - 20% of the smuggled substance. The preparation of the drugs involved wrapping them in brown packing tape, thought to be a common practice of drug smugglers.
3. The drug smuggler would then take the wrapped package and conceals it in the vehicle. This requires him to handle the vehicle keys and open the front driver door and then either the boot or one of the rear passenger door.
4. The keys are passed to the proxy who then simulates driving the vehicle, opening and closing the drivers door and the boot and one of the passenger doors, not necessarily the same door as was handled by the packer.
5. The proxy would then put his fingerprints on a sheet of aluminium foil to simulate a secondary transfer surface. The prints of the drug packer would also be taken for reference.
6. The vehicle would then be sampled.

This scenario was undertaken for each of the following contraband narcotics:

- Cannabis resin
- Herbal cannabis.
- Amphetamine sulphate.
- Ecstacy (MDMA).
- Heroin.
- Cocaine.

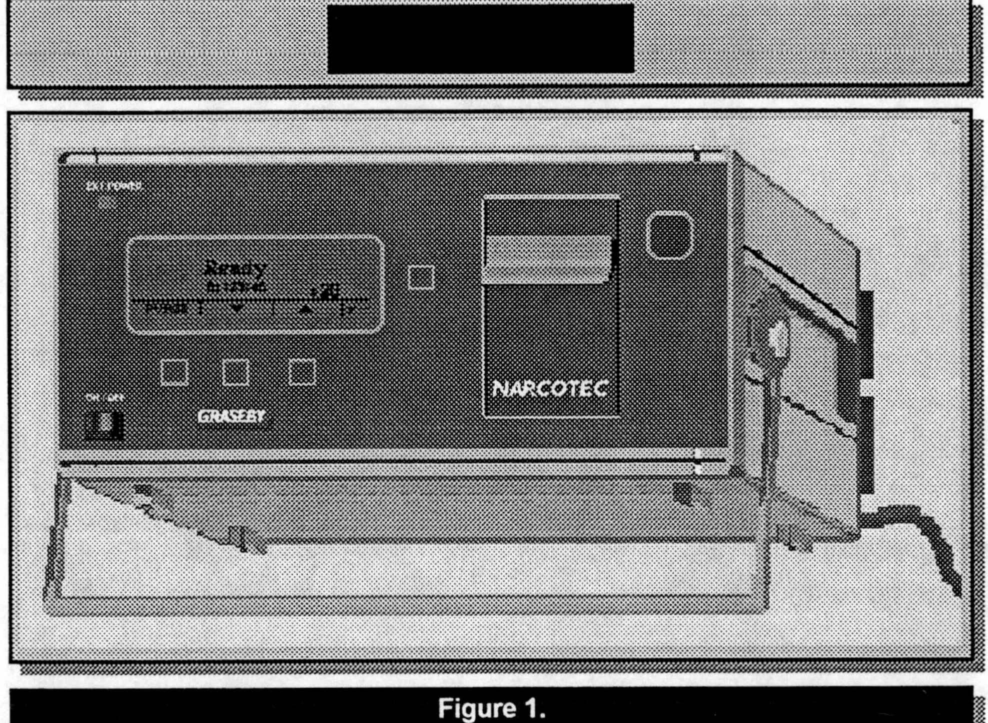

Figure 1.
DETECTOR FOR ILLICIT DRUGS

Detection System.

The entire detection system used during the trial comprised of several components.
a) The Graseby NARCOTEC™ detection unit, which uses thermal desorption hot IMS. This consists of a heated IMS cell of a ceramic based construction (maintained at 200C) with a hot air jet used to desorb the sampled material off the sample wipe. A pictorial representation of the system and the system pneumatics are shown in fig. 1 and 2. The system weighs 17kg and is man-portable. This unit provides an alarm indictor, response level indicator and drug identifier.
b) Sampling wipes, consisting of a 25mm dia. cotton filter on a card carrier, the wipe inserts directly into sample port on detector.
c) An associated laptop PC and software provides data storage facility for record keeping, spectral display, and detection parameter display.
The entire analysis cycle takes 15s consisting of 12 analysis spectra, each of these spectra comprising of an average of 32 IMS spectra.

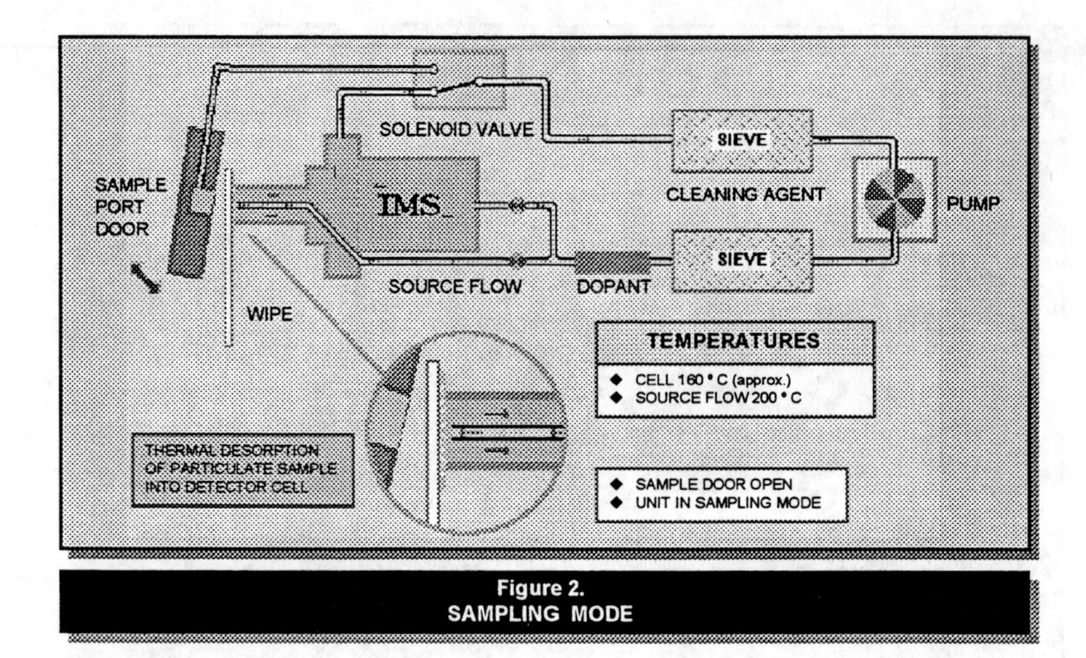

Figure 2.
SAMPLING MODE

Experimental Method

Great care was taken at all stages to ensure that the search area and the vehicles were initially clean, and that there was no cross contamination of the people involved from the surroundings. All drugs were initially removed from their storage packing in an adjacent room and transferred into the test room wrapped with only one layer of packing, often a polythene bag. The package preparation area was always covered with clean paper and tested before packing. All the vehicle surfaces, and hands of all personnel participating in the test room were checked before each concealment was made.

A simulated concealment was made by the drug packer opening the car driver's door using the key, and then opening the boot and/or one of the passenger doors. The drug package was not placed in the vehicle as this was judged to have little effect on the traces on the exterior surfaces of the vehicle.

All sampling was done by simple wiping . First, each car was wiped using a single wipe: the car door handles and contact surfaces were wiped starting with the front passenger door, moving round to the boot and onto the driver's door last. The wiping process followed closely that which might be performed in a field deployment. Each wipe was analysed directly following the sampling. A full hand print of the drug packer and the proxy was made on to aluminium foil and then sampled with one wipe. The foil was replaced between each sample and check for contamination prior to the prints being deposited.

The packing method for each of the drugs was:

- **Cannabis resin.** This package was made by wrapping 12 kilos of cannabis resin.

The drug was already wrapped in a light brown tape,and to simulate typical packing damage, was slightly broken.

- **Cannabis herbal.** Large quantities of herbal were not available; however a small quantity typical of that smuggled for personal usage was obtained. This was handled by the packer and the concealment made as before.
- **Heroin.** A small pack of street heroin was available the outside of the small polythene package was handled by the packer. The concealment made as before.
- **Amphetamine.** Several polythene bags of amphetamine tablets were taped with heavy masking tape.
- **Ecstacy (MDMA).** The package was made by wrapping bags of approx. 3kg of ecstacy tablets in tape, and the concealment made as before.
- **Cocaine.** The package again was a polythene bag of cocaine powder wrapped in tape and the concealment made as before.

Wipes were taken off:

- the contact surfaces of the vehicle
- the hands of the proxy
- the fingerprints of the proxy.
- the fingerprints of the packer
- the hands of the packer.

The response levels in the target windows of the IMS spectra were recorded and these levels compared to the response from levels recorded using standard solution tests, where calibrated amounts of the different drugs are deposited directly onto the wipes and a corresponding response level determined. The response characteristics of an IMS detector are such that a linear response is usually only possible over 2 orders of magnitude and therefore only where the measured signal is in the linear region of the response curve can a linear extrapolation of the mass to signal amplitude be made. Using this information estimations of the mass of drug material transferred to the wipes were made. While this type of experiment is quite simplistic, the order of magnitude information is quite useful when determining the possible successes of applying this technology to the drug targeting problem.

Results.

The result for each of the different drug materials is presented below.

- **Cannabis resin.** From this package no traces of tetahydrocannabinol THC (the targeted component) were found on either the car, the hands of the proxy, the secondary surfaces from either the packer or the proxy. It appeared unlikely that any detectable levels of THC were transferred to the packer from the resin. The package did appear quite dry on the outside and the THC, an oily material, was probably not present on the surface of the blocks of drug in sufficient quantities for the necessary transfer processes to take place. A significant level of THC was however found on the surfaces of the table where the package had been prepared. Figure 3. shows the spectrum of this response. This response probably comes from "flakes" of the resin picked up from the surface of the table.

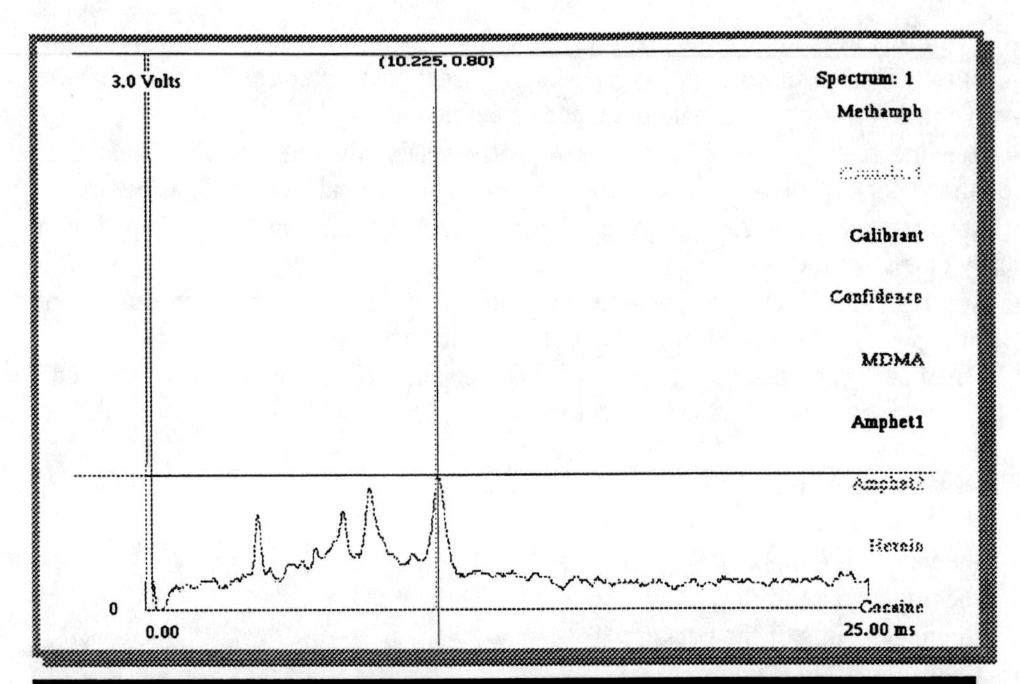

Figure 3.
WIPE OF TABLE SURFACE USED TO PREPARE CANNABIS RESIN PACKAGE.

- **Herbal Cannabis.** Traces of the THC were found on the car and a clear response is seen in the IMS spectrum, approx. 0.25V (measured above local background) (see figure 4). This level of response is near the alarm threshold. The response from the hands of the proxy was smaller, approx. 0.1V (see figure 5). The prints from the hands of the proxy on to the foil, however showed a larger response: about 0.2V. (see figure 6). The prints from the hands of the packer however showed a large response (1.25V)(see figure 7). The facility of the system to detect cannabis is obviously dependent on the age and state of the initial drug. Although there is evidence to suggest that traces can be found, at near the detection limit of the system, on vehicles and secondary transfer surfaces from both the packer and the proxy.

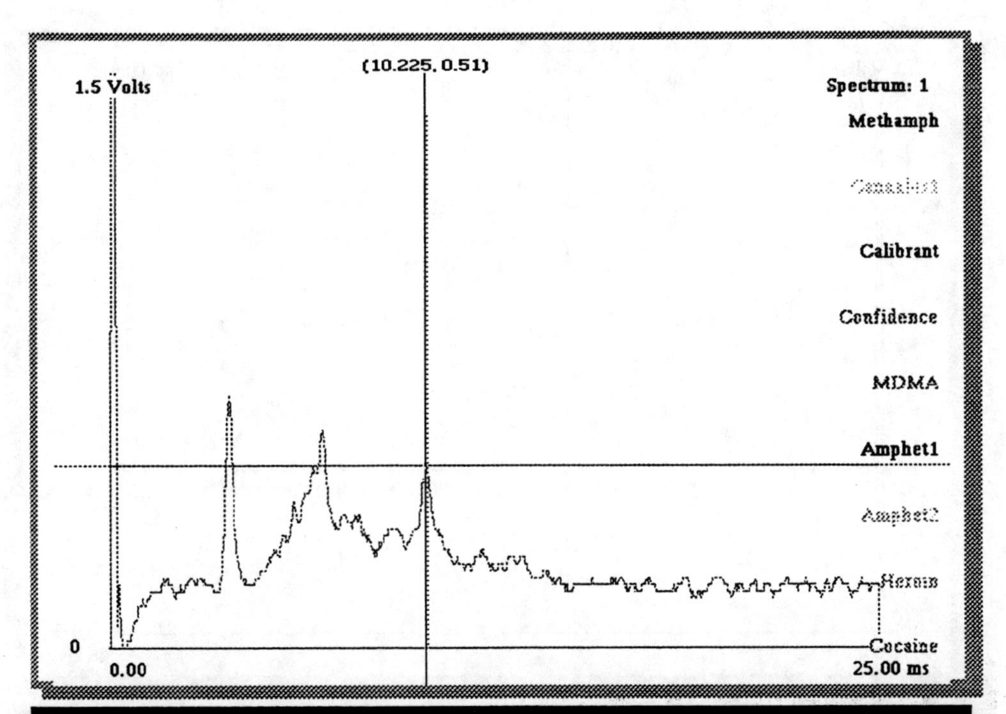

Figure 4.
WIPE OF VEHICLE CONTACT SURFACES FOLLOWING CONCEALMENT OF HERBAL CANNABIS.

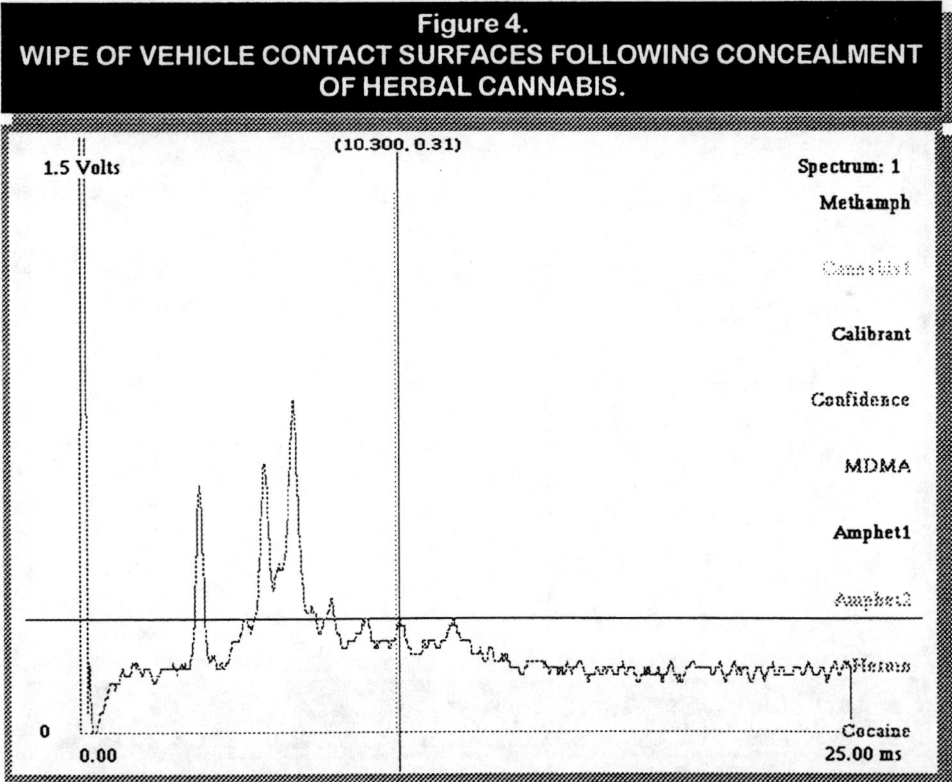

Figure 5.
WIPE DIRECT FROM HANDS OF PROXY (CAR DRIVER). CAR HAS HERBAL CANNABIS CONCEALMENT.

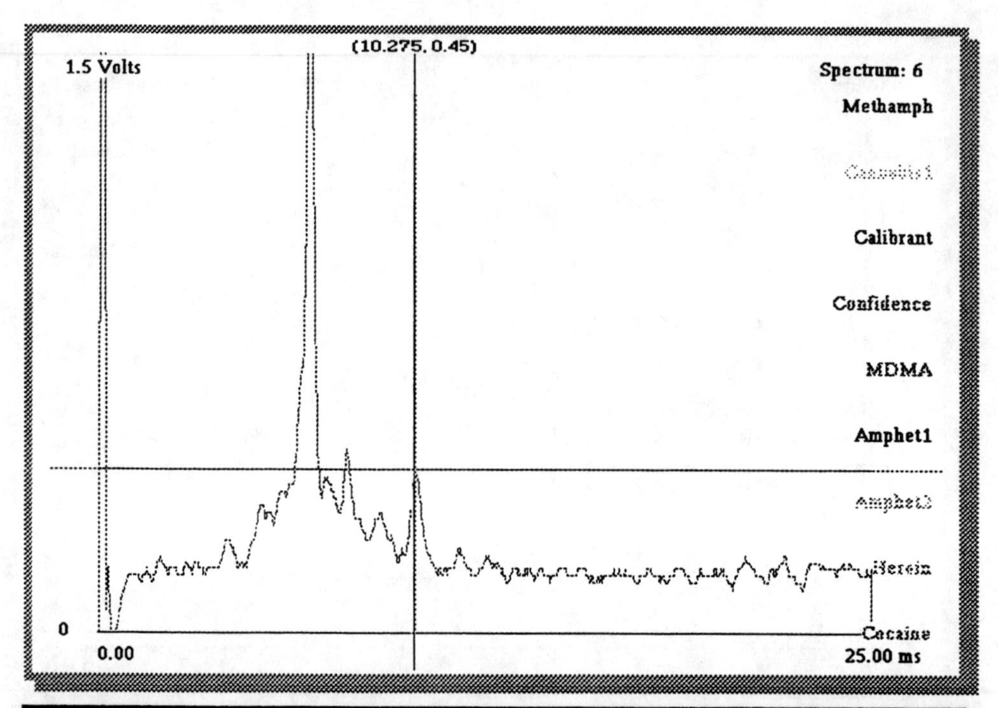

Figure 6.
WIPE FROM FOIL WITH FINGERPRINTS OF PROXY.

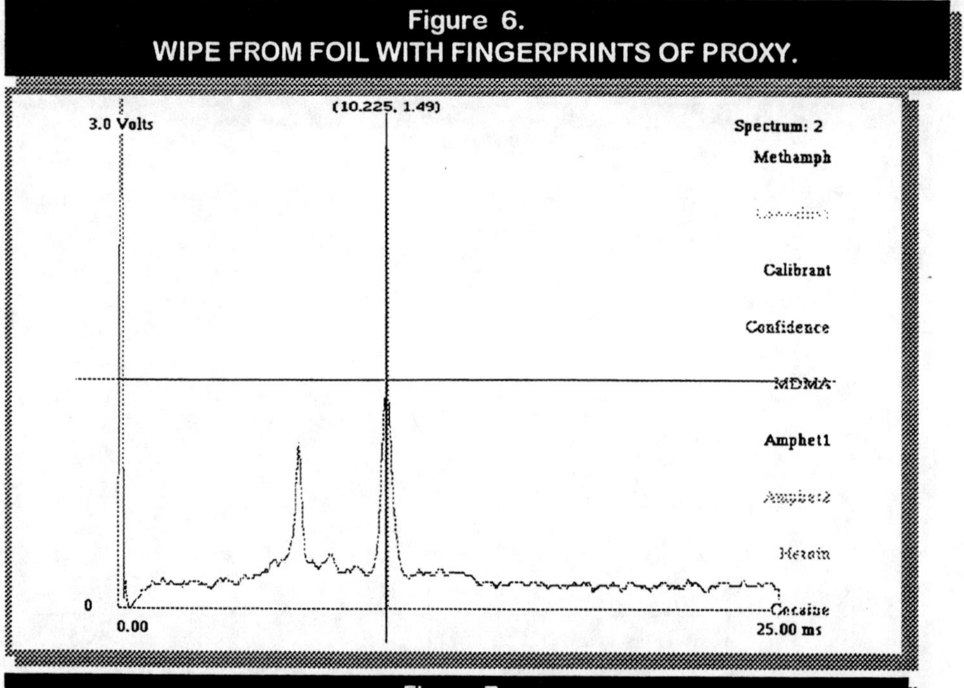

Figure 7.
WIPE FROM FOIL WITH FINGERPRINTS OF HERBAL CANNABIS DRUG PACKER.

- **Heroin.** A large peak from the heroin from the car wipe was shown (1.0V) (see figure 8). The prints on foil from the hands of the proxy gave a response of 0.6V. This is obvious evidence of the transfer of drug material to the car, the proxy, and any secondary surfaces at levels above the detection limit of the system (figure 9). Other distinct peaks are present in the spectrum probably reflecting the cut of the heroin as this was a "street" sample. It is not ussually possible to use the cutting agents as indicators of the drug as these would not be consistent in all concealments..

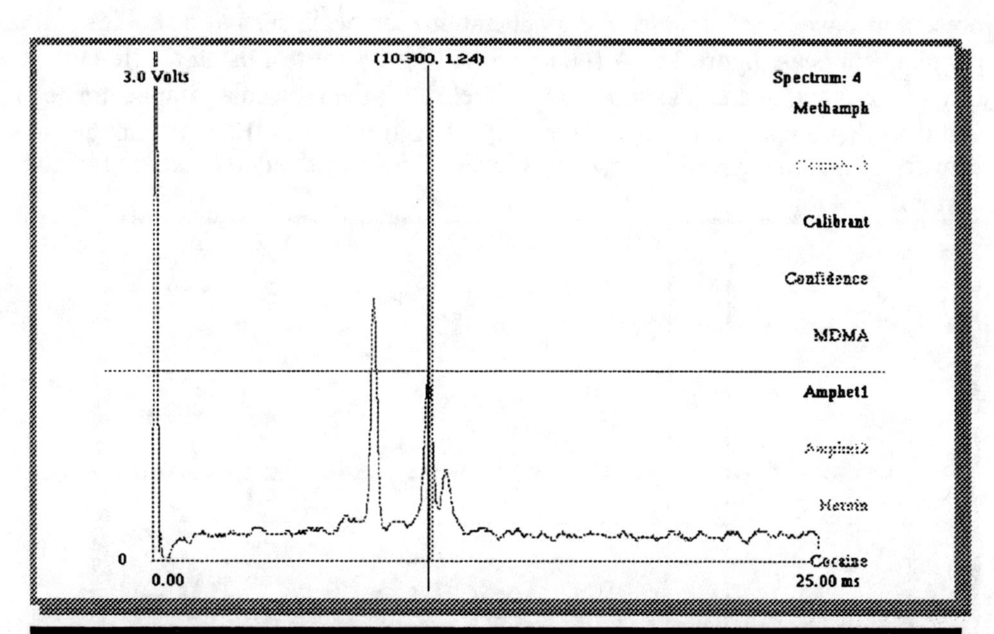

Figure 8.
WIPE FROM VEHICLE CONTACT SURFACES WITH CONCEALED HEROIN SAMPLE.

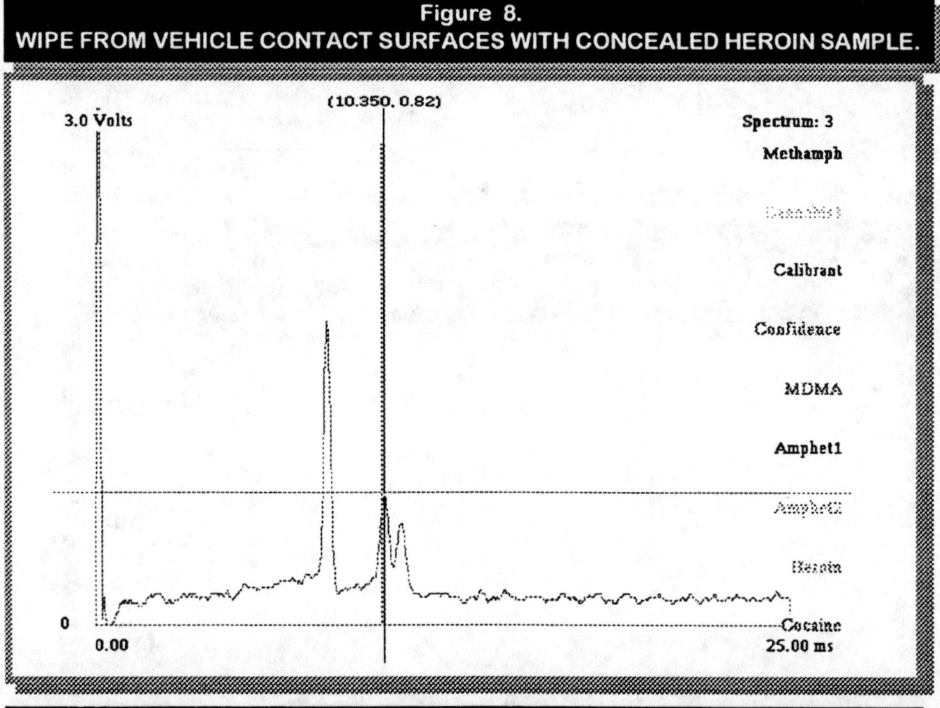

Figure 9.
WIPE FROM PRINTS OF PROXY ON FOIL, HEROIN CONCEALMENT

- **Amphetamine.** Off the wipe from the car a large peak was detected at 1.30V (figure 10); although this was not the standard amphetamine target peak. This peak has been detected in other cases from street amphetamine; it does not correspond to the molecular amphetamine ion and is not present when a pure amphetamine standard is used. This peak was traced back to the outside of the packaged drug. It was also found off the foil with both the prints of the packer and the driver;

present also was a small molecular amphetamine ion peak, seen as a shoulder on a contaminant peak, figure 11. A repeat of the finger print test using a different original package of amphetamine showed peaks for the molecular amphetamine and the other amphetamine peak at similar intensities (Figure 12.). Again the transfer of material gave responses at a level well above the detection limit of the system.

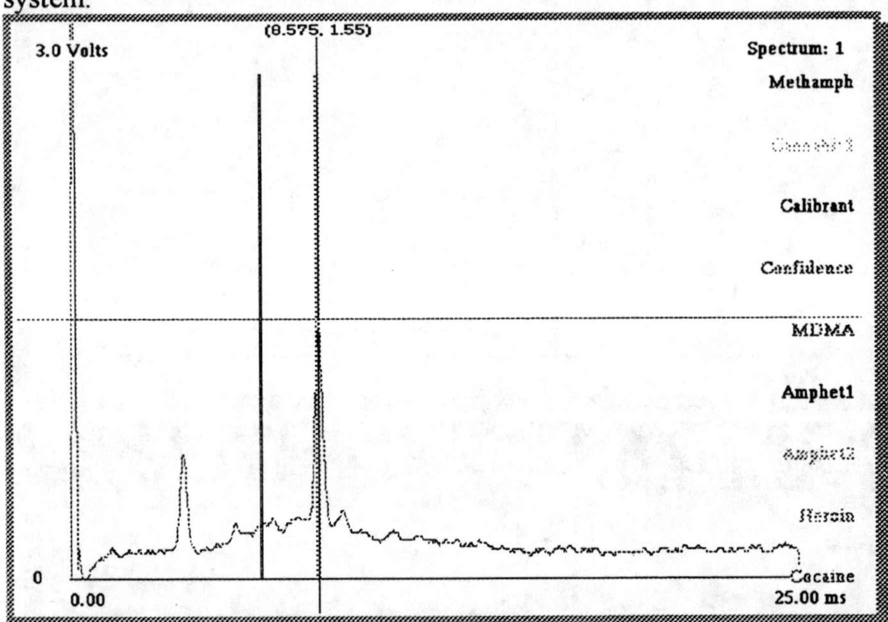

Figure 10.
WIPES FROM CONTACT SURFACE OF VEHICLE FOLLOWING AMPHETAMINE CONCEALMENT.

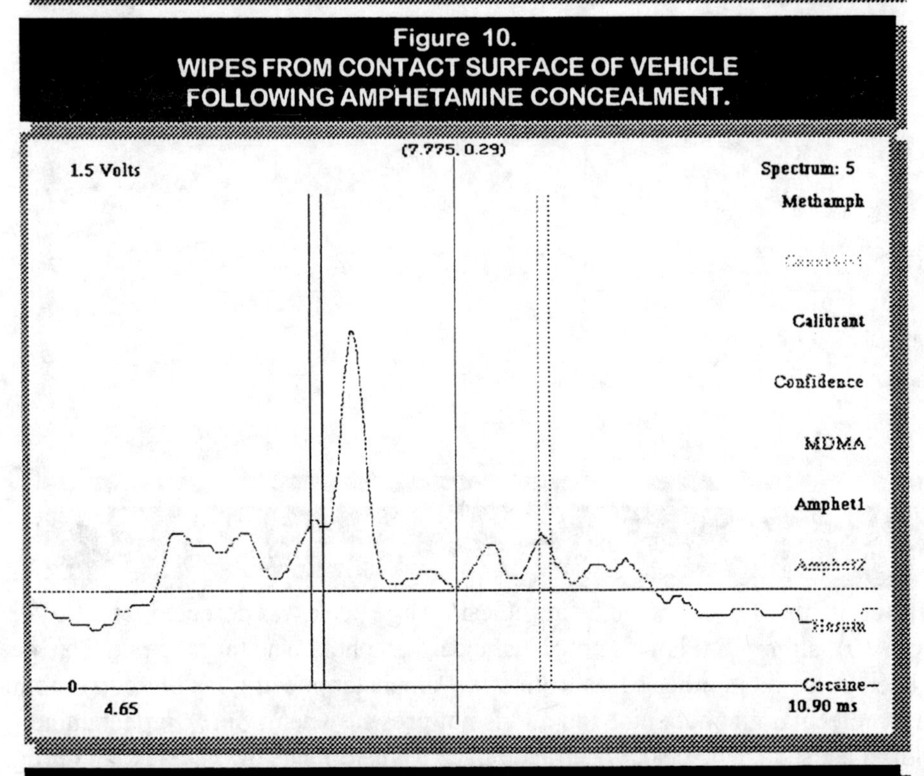

Figure 11.
WIPE FROM PRINTS OF PROXY ON FOIL, SHOWING BOTH AMPHET PEAKS. PEAK 1 AS A SHOULDER ON INTERFERENT.

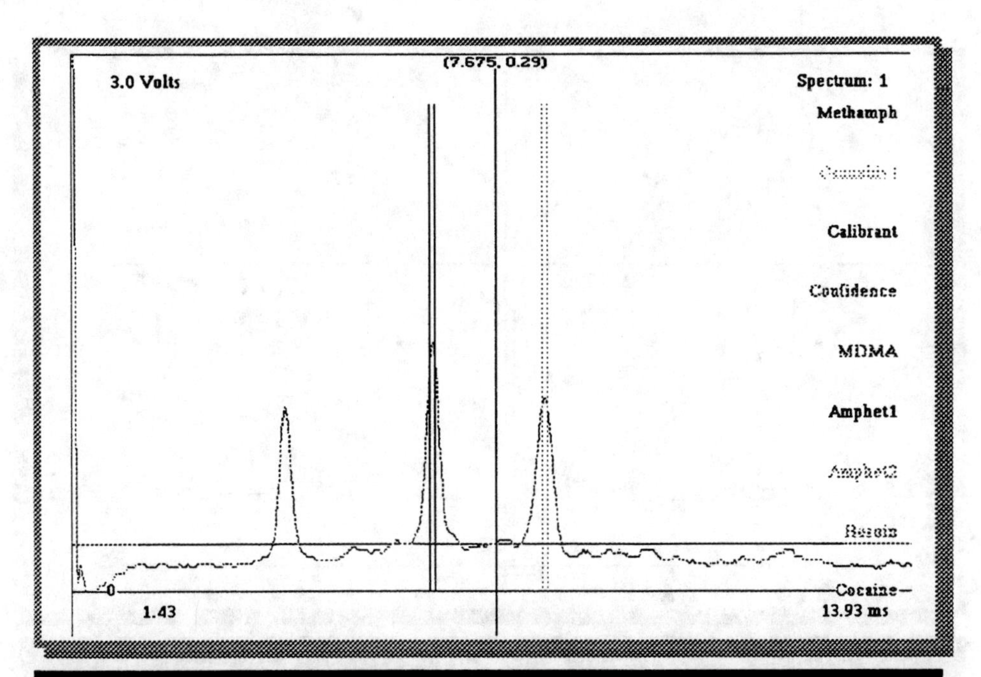

Figure 12.
WIPE FROM PRINTS OF PACKAGER OF SECOND AMPHETAMINE SAMPLE SHOWING BOTH TARGET PEAKS.

- **Ecstacy (MDMA).** While a large peak was found in the spectrum from the car wipe, at the beginning of the thermal desorption cycle the peak was outside the target window and moved toward the centre of the window by the completion of the analysis cycle (see figures 13 & 14). It is suggested that this was due to effects on the ion chemistry of the product ion by the interferents from background dirt off the car. Probably with high concentrations of moisture in the source region the equilibrium in the combinations of water with the molecular ion shifts so that larger molecules are created. As the analysis cycle proceeds then the moisture level in the source region falls and the equilibrium shifts back to the norm expected for a "dry" system.

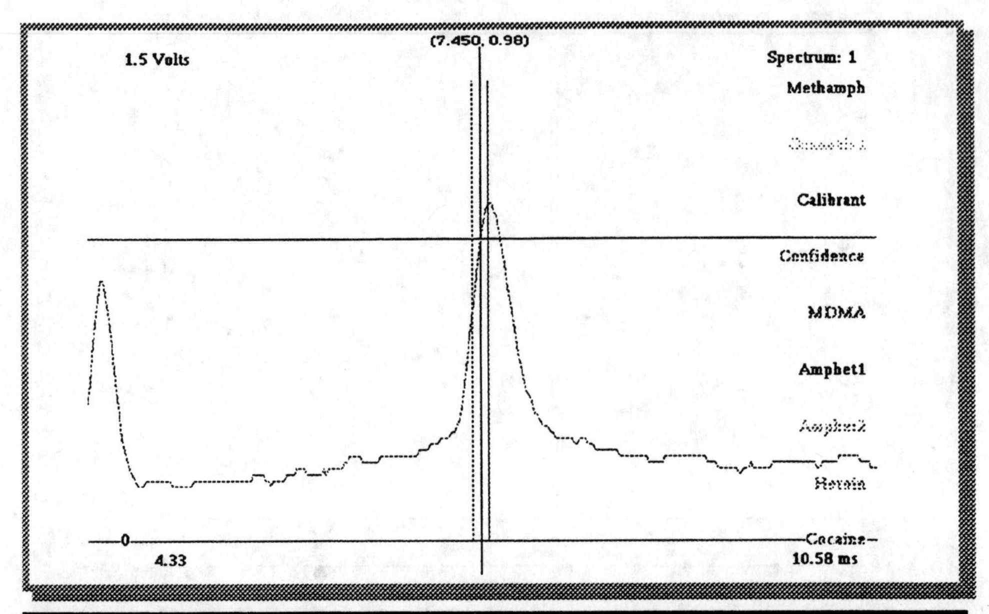

Figure 13.
WIPE FROM VEHICLE CONTACT SURFACES WITH ECSTACY CONCEALMENT.
MDMA PEAK AT BEGINNING OF ANALYSIS CYCLE, OUTSIDE WINDOW ON LHS.

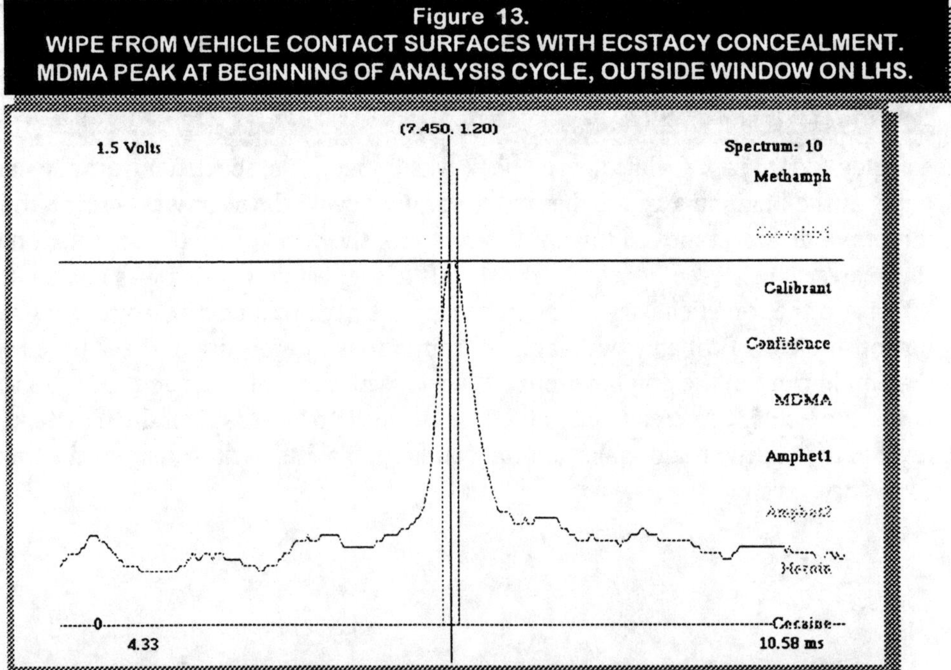

Figure 14.
WIPE FROM VEHICLE CONTACT SURFACES WITH ECSTACY CONCEALMENT.
MDMA PEAK AT END OF ANALYSIS CYCLE, CENTRE OF WINDOW

The tests of the prints on the foils gave large target peaks and again there would be background interferent from skin oils here. This effect can also be seen in the asymmetric nature of the MDMA molecular ion peak showing a tail on the high drift time side of the peak. These effects require that for reliable MDMA detection the peak width criteria and peak targeting are adjusted from that expected from lab tests. The transfer of drug material again was at levels well above the detection limit of the system.

- **Cocaine.** Very large response peaks were found at all stages of transfer. Off the car it was at a level of 3.5V. The transfer of the cocaine is very strong and the response of the system to cocaine is very strong. The measured responses off all surfaces were at or near saturation level for the detector, typically, >3.0V. Cocaine was still detected at high levels (1.2V) off the hands of the proxy, even after washing. A car brought into the area where the packaging was being done, was covered in easily measurable trace levels of cocaine even though it was not used for drug concealment. The car for the main concealment test had to be kept outside the packing area. When the car was dowsed in water and a wipe taken so that the cotton filter was near saturated with water, then the response was still a strong peak. The peak however drifted across the IMS spectrum converging on the cocaine target window, figure 15. This drift is probably due to high moisture levels in the drift region.

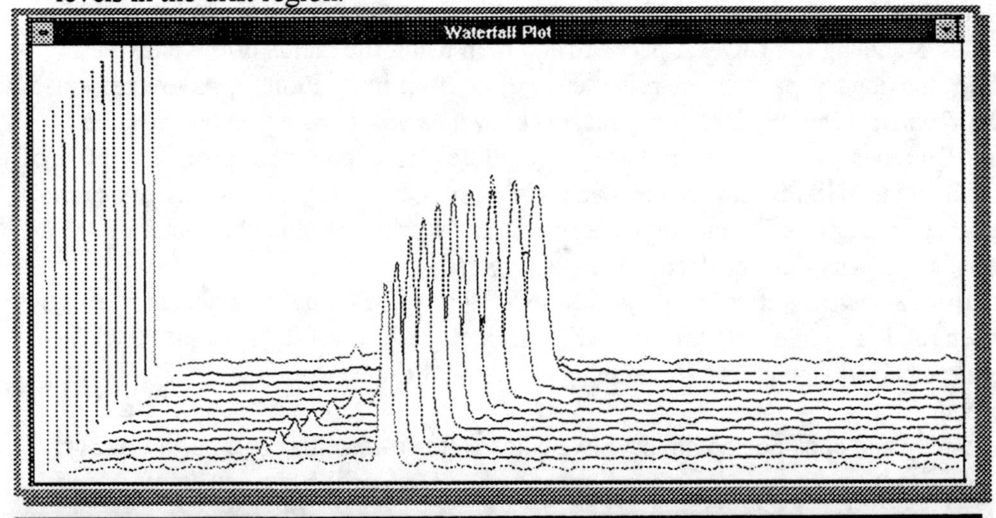

Figure 15.
VARIATION IN POSITION OF COCAINE PEAK DURING ANALYSIS CYCLE FOR VERY WET WIPE.

Conclusions

1. With the exception of cannabis the transfer of material from all drugs tested to primary (the vehicle) and secondary (the prints of the proxy) surfaces is strong. For cocaine this is exceptionally so.
2. For cannabis there is some weak transfer of the THC but it is strongly dependent on the nature of the original material. Evidence suggests that transfer from resin is poorer than herbal.
3. For cannabis the trace deposits of THC on primary contact surfaces depend on the nature of the primary material, for herbal they could be upto 500ng however the response of the detection system, including the wiping efficiency is thought to be lower for cannabis targeting.
4. The wiping routine deployed and the efficiency of detection of the system is sufficient to detect the traces.

5. For the hard drugs trace levels of drug removed by wiping from primary contact surfaces is typically in the range 500 to 1000ng (heroin, cocaine, ecstacy, amphetamine,). For these materials the wiping efficiency is estimated at between 25-50%, Therefore the trace levels present on the contact surfaces is typically in the few microgram range.
6. The peak measurement routines in the system have to be designed to match the nature of the response from field applications. The detection alogorithms developed from this study for the NARCOTEC have proved very successful in field deployment.

Additional tests

An additional test was performed to study the persistence of the transfer of the drug from the packer to the primary transfer surface. This is a test similar to those performed with explosive materials. A repeated print test was performed using MDMA. Handling the MDMA package and then using the hands of the packer (gloved) making a repeat thumb print series up to 36 prints. Print 36 gave a response of 0.4V with a generally increasing response level as you progressed back up the series. By repeat wiping of a print area it could be estimated that approx. 25 - 50% of the available material is removed by each wiping process. Trying to transfer material by rubbing an ungloved thumb on an untested early print, produced transfer of material but at a significantly reduced level, typically 5%.
Tests from a passport after handling "SKUNK" a form of herbal cannabis increasingly found in the U.K, gave a significant THC response peak, level 0.8V, higher than the other type of herbal used, while sampling directly off the hands a level of 1.5V was recorded.
The ecstacy and cocaine contaminated vehicles were re-tested after a period of 2hrs and significant traces were still found, greater than 50% of the previously measured level.
After 3 weeks of normal vehicle usage the cocaine contaminated vehicle was re-tested and significant traces still found, typically 10% of the previously measure value.

Some Current Activities in Trace Explosives Detection A the FAA Technical Center

Frank T. Fox, Ph.D.
Federal Aviation Administration
William J. Hughes Technical Center
Atlantic City International Airport, NJ 08405

The focus of trace explosives detection activities at the Technical Center has been to develop standards which are accurate and repeatable, and to use these standards to certify commercial instruments. A feature of plastics explosives which had to be retained in the standard was their tendency to adhere to surfaces. Part of this property is attributed to the polymer coating which surrounds the explosives crystals. The coating is applied during the manufacturing process of C4 (RDX), Semtex H (RDX plus PETN), and Deta Sheet (PETN). The crystals in these three explosives are coated with Polyisobutylene, styrene-butadiene, and nitrocellulose respectively. Simply using aliquots of a solution of these explosives in an organic solvent does not work because the solvent removes the polymer coating.

For the purposes of the certification process retention of the character of the original explosive is important because the traces detected by inadvertent contamination of the outer surfaces of a container or clothing are from the original explosive used to make the bomb. The sampling characteristics, that is, the ease of removal of explosive from the test surface must be the same in the certification test object as in the "naturally" expected sample.

We have now developed a method for the consistent, accurate production of deposits of plastic explosives which have the same sampling characteristics, namely "stickiness", as the bulk explosive. The method works well with the major commercial and military plastics. The technique takes advantage of the fact that plastic explosives are manufactured as crystals of explosive which are stabilized and protected by coating with plastic, such as polyisobutylene or styrene-butadiene. The result is encapsulated crystals which are made into the final product by pressing the particles together into bricks or sheets, along with a little oil or other agent to aid flexing. Normal production of a fingerprint occurs by touching the bulk explosive and removing some of the coated particles mechanically from the bulk because they stick to the skin. The particles are then transferred to the next object or surface touched.

Our method takes advantage of the encapsulation resulting from the manufacturing process. The adhesion forces between the coated particle in the compressed bulk are overcome mechanically. A small piece of plastic explosive is agitated with water causing particles of the coated explosive to be simply shaken loose from the bulk and suspended in the water. By analogy, touching the bulk causes mechanical removal of particles due to pressure and adhesion.

The explosive is not dissolved by water agitation process, but particles from the bulk are merely suspended by gentle mechanical means. Variations on the agitation include stirring, sonication and shaking. Once the explosive is suspended in its normal form of coated crystals, the actual amount is determined by usual methods including HPLC, GC and MS. Dilution is then made to an appropriate level.

In order to prepare a deposit consisting of the desired amount of explosive, an aliquot of the diluted and quantified suspension is pipetted onto the surface of a test object and the water evaporated. The dried residue consists of plastic coated explosive crystals in the original state as would occur with a natural fingerprint or smear. The amount of dried residue deposited can easily be varied by simply controlling the degree of dilution of the original suspension of explosive particles. We have prepared reproducible deposits ranging from high micrograms to low nanograms. The concentration of the deposit is verified by back extraction of a test surface and quantitation of the extract. Particle size of each preparation is verified by scanning electron microscopy. Consistent particle size range in the deposit was our goal in the preparation of a standard. Comparability of the particle size range of the deposit to actual fingerprints was also considered important. An unusable variation in the amount deposited would be large if large particles appeared. Consistent analysis with RSD's of less than ten percent were obtained on back extraction and analysis of the deposits. In addition, a close comparison of particle size between a fingerprint and a deposit of the suspension of C-4 was observed.

The controlled deposition of several plastic explosives on some typical carry-on electronic items was used on commercial instruments at the 1995 ICAO meeting of the "AD HOC Group of Specialists on the Detection of Explosives", held at the FAA Technical Center in Atlantic City, NJ. Six commercial instruments were tested over several days using the standard sampling technique proposed by the trace detection instrument manufacturer. Participants in the exercise included manufacturer representatives and an international group of persons experienced in trace explosives detection .

One of the concerns that has developed regarding TDS systems is their vulnerability toward interfering compounds which could mask a positive alarm signal for an explosive compound.

We examined two different technologies; A Thermal Energy Activation (TEA) system, the EGIS 3000, and an Ion Mobility Spectroscopy system (IMS), the Barringer 400 Ionscan. We found that both systems are effected by obscurants. With IMS, the process is likely to be that the reactant ion formed undergoes a side reaction with the obscurant in the reaction region of the drift tube, sometimes to the exclusion of forming the desired product with the explosive, hence no alarm signal; A rough idea of this interaction can be obtained by comparing the relative concentrations necessary to achieve obscuration (see

Tables 5 and 6). For some obscurants it is possible to observe this process by comparing plasmagrams of explosives tested in the presence of sequentially higher concentrations of a given obscurant. We have seen that the marker peaks for the explosive continuously decrease while the peak for the obscurant increases. A second obscuration process that we have seen is due to peak shifting of the explosive marker peaks. Compound "S" was notable for this and uniformly shifted the peaks of interest out of their identification time slots leading to a "no alarm" situation in the presence of explosives. Again, this is easily observed from the plasmagrams. The process for obscuration of a TEA system, such as the EGIS 3000, is not as clear.

TESTING FOR OBSCURATION

1). Select candidate obscurant.

2). Develop a compatible solvent system and prepare concentrated obscurant.

3). Screen materials on TDS without explosive, if a positive alarm is found eliminate material from further testing.

4). **PHASE ONE;** Test obscurant candidate in presence of explosive using largeamount of obscurant and direct application to system's desorbtion system. if 40% or more tests do not alarm, retain for further testing;.

5). **PHASE TWO;** Test material against 3 X MAL of explosive, using serial dilutions looking for the lowest level of obscurant that will reliably negate alarm. This is the obscuration level. When this is found, reduce this level to the highest amount of material that will not obscure the alarm, 20 out of 20 times. This is the challenge level. This is done on the system's sample desorbtion tab as well.

6). **PHASE THREE;** To test the overall pattern of obscuration for the system, develop system minimum alarm level (SMAL) of explosive for a given TDS unit and apply the procedure of step 5, sampling from an external test object using the manufacturer's sampling procedure.

7). The data collected will give a series of compounds and dilutions that can be used to compare different TDS instruments for susceptibility to obscuration of a certain material in the presence of a particular explosive.

This work represents an initial study on the characteristics of obscuration. For RDX, Phase Two produced four materials with varying degrees of obscuration on the Barringer 400, Compounds "S", "G", "Q", "D"(Fig. 1). The EGIS 3000 yielded two candidates for RDX obscuration, Compounds "G" and "Q" (Fig. 3). The EGIS 3000 system also yielded two PhaseTwo PETN candidates, Compounds "T" and "Q".

Phase Three studies were done only on the Barringer 400 with C-4 as the explosive tested (applied in suspension). In this study, Compounds "S", "D", and "Q" were found to yield consistent obscuration of alarm response. This is illustrated in Figure 2.

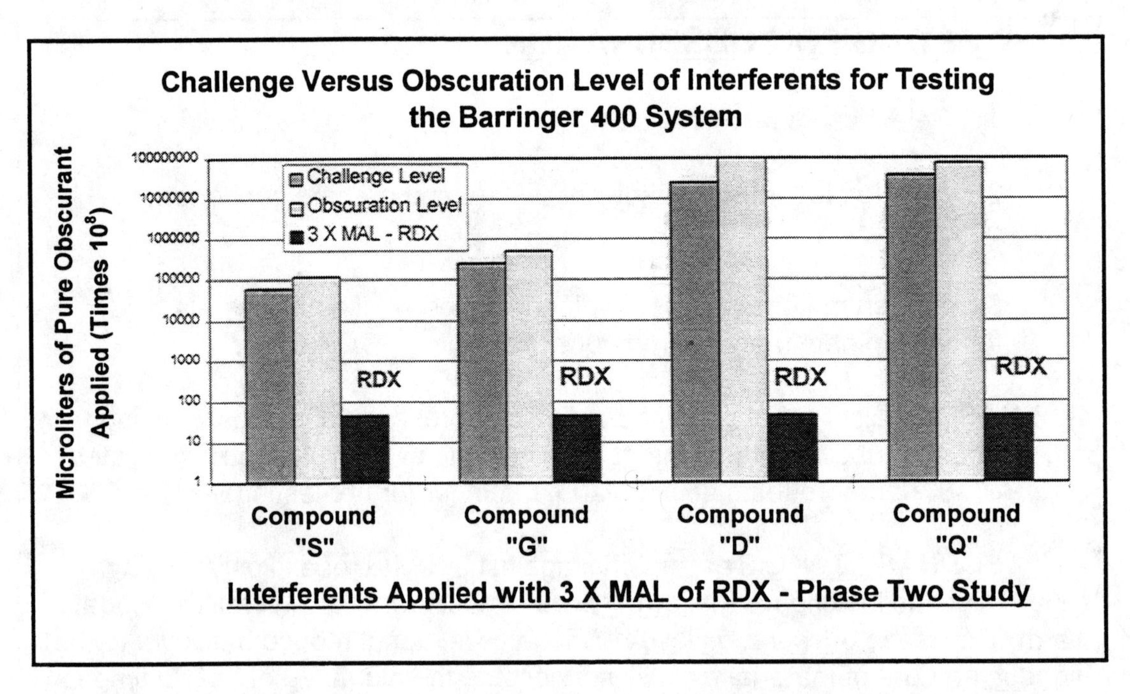

TABLE 1). Comparison of challenge and obscuration levels of obscurants for RDX on the Barringer 400 system. Phase Two trials.

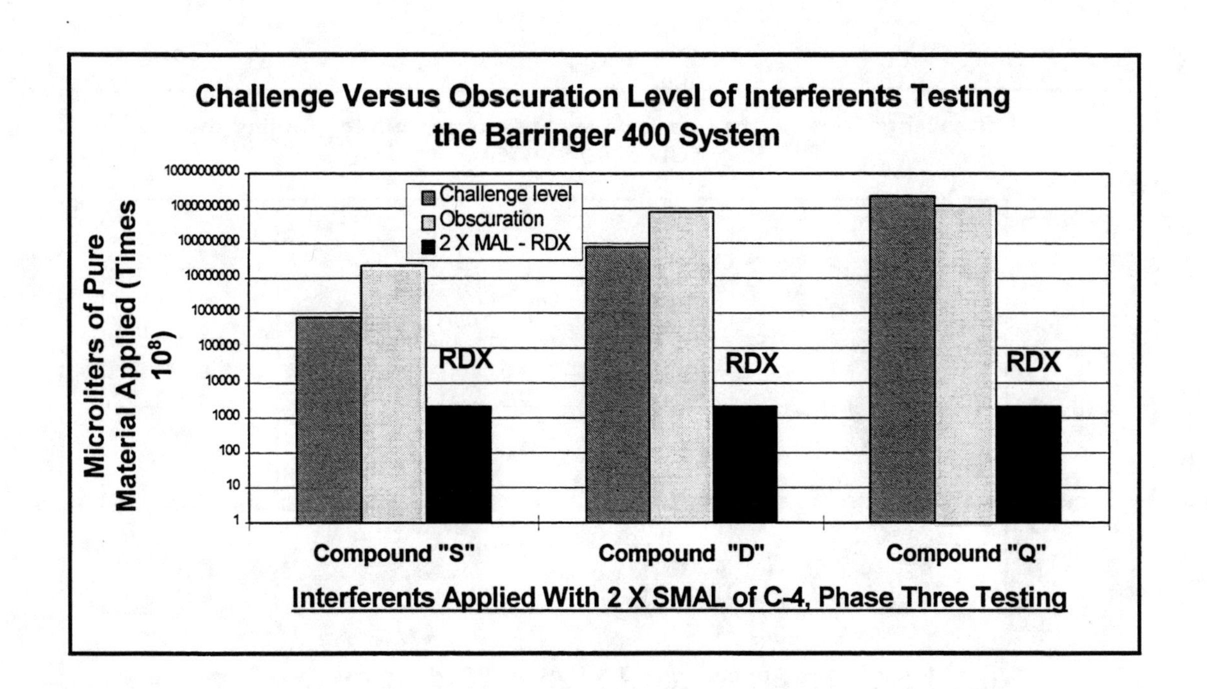

TABLE 2). Comparison of challenge and obscuration levels of obscurants for RDX on the Barringer 400 system. Phase Three trials.

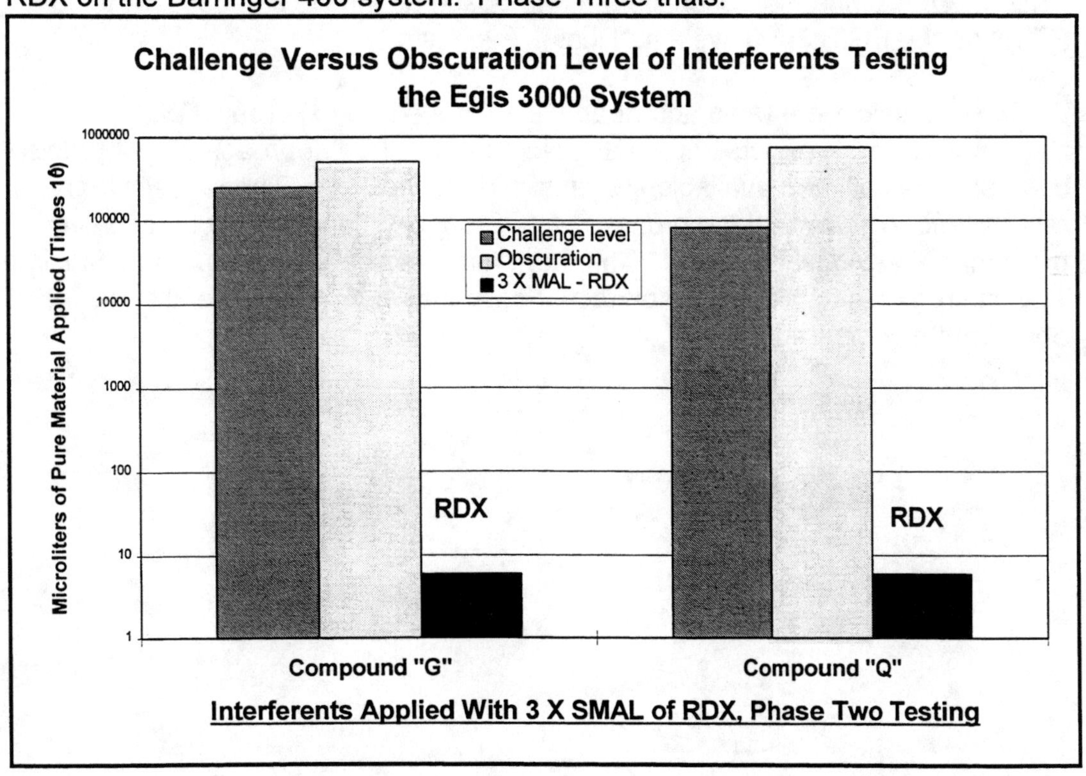

TABLE 3). Comparison of challenge and obscuration levels of obscurants for RDX on the EGIS 3000 system. Phase Two trials.

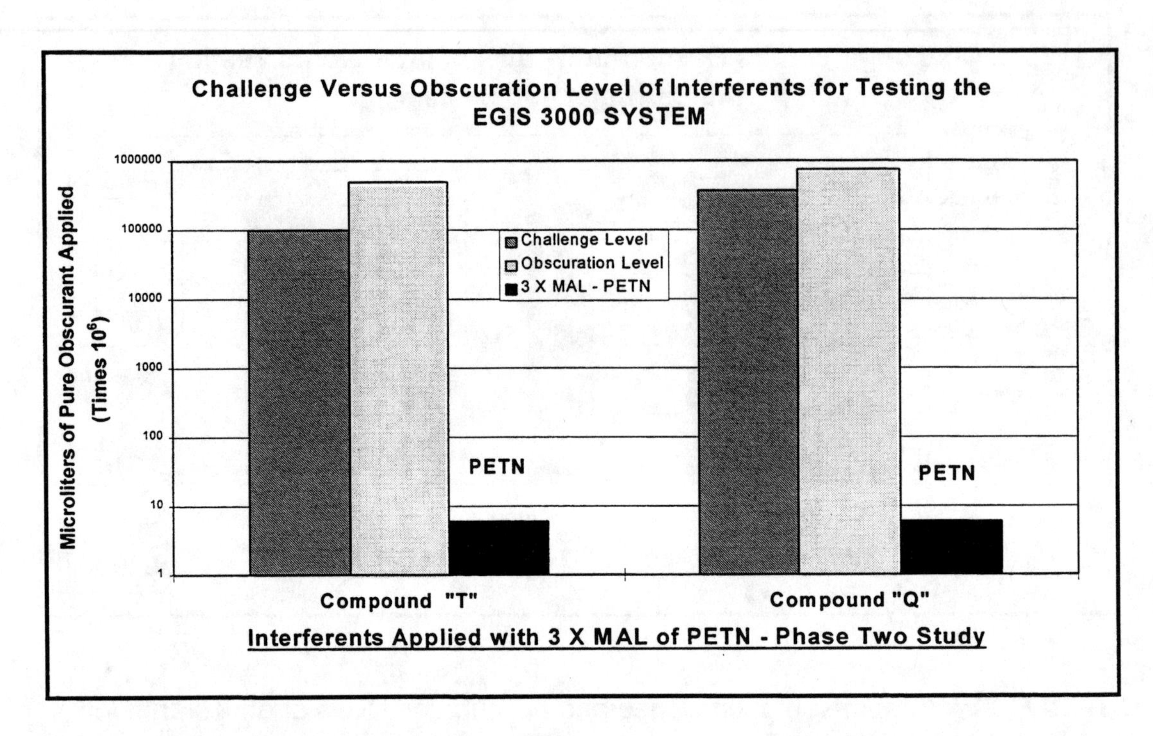

TABLE 4). Comparison of challenge and obscuration levels of obscurants for PETN on the EGIS 3000 system. Phase Two trials.

We have developed a rapid and quantitative procedure to screen TDS instruments for the effects of potential obscurants. This approach allows precise comparisons between available systems and technologies. These preliminary results tend to show that the process of TDS obscuration may not be as severe a problem as previously believed. This work may also be a step toward optimizing TDS instruments to reduce the effects of obscuration through careful analyses of obscurant plasmagrams.

NDTA narcotics standard development

S.J. Ulvick, J. Cui, T.D. Kunz
Houston Advanced Research Center
4800 Research Forest Drive, The Woodlands, Texas 77381

D.E. Hoglund
United States Customs Service, Applied Technology Division
1301 Constitution Avenue, NW #6212, Washington, DC 20229

P. Pilon, A.H. Lawrence
Revenue Canada, Laboratory and Scientific Services Directorate
79 Bentley Avenue, Ottawa, Ontario, Canada K1A OL5

G. Drolet
Revenue Canada, Customs Operations Branch
191 Laurier Avenue West, Ottawa, Ontario, Canada K1A OL5

C.W. Su
United States Coast Guard, Research and Development Center
1082 Shennecossett Road, Groton, Connecticut 06340

S.W. Rigdon
Analysis and Technology, Incorporated
258 Bank Street, New London, Connecticut 06320

J.C. Demirgian
Argonne National Laboratory
9700 South Cass Avenue, Argonne, Illinois 60439

P. Shier
Kentco, Incorporated
158 Magnolia Road, Sterling, Virginia 20164

ABSTRACT

The Narcotics Detection Technology Assessment (NDTA) program is a series of studies conducted to evaluate illicit substance detection devices. The ability to effectively detect cocaine and heroin particles is directly related to the efficiency of a detection device's sample collection design. The NDTA tests are therefore structured to require sampling of narcotics from a surface. Test standards are required which permit subnanogram to microgram quantities of narcotic to be dispensed onto a target surface for sampling. Optimally, the standard should not adversely affect the performance of the device under test.

The NDTA test team has developed and experimentally characterized solution-deposited substrate standards, solution-deposited substrate-free standards, vapor-deposited standards, suspension standards, and dry mix standards. A variety of substrates and dry-mix fillers have been evaluated, including sand, fullerenes (bucky balls), copper powder, nickel powder, silanized glass beads, monel powder, cellulose powder, silver-coated nickel powder, pulverized paper, and aluminum. Suspension standards were explored with a variety of liquids. The narcotic standards with the best performance were found to be dry mixes of cocaine with silver-coated nickel powder, and dry mixes of heroin with silanized glass beads.

Keywords: NDTA, standards, cocaine, heroin, narcotics, mixes, drugs

1. INTRODUCTION

The United States Customs Service (USCS) Applied Technology Division (ATD) of the United States Treasury Department is conducting independent technology evaluation studies of illicit substance detection devices[1-4]. These studies, named the Narcotics Detection Technology Assessment (NDTA) Program, are sponsored by the Office of National Drug Control Policy (ONDCP), Counterdrug Technology Assessment Center (CTAC). The USCS directed NDTA studies are designed to evaluate both commercially available and prototype chemical based detection systems in laboratory and field simulation scenarios.

The primary objective of the NDTA program is to *quantitatively* evaluate the detection performance of chemical based narcotics detection equipment. Specifically, these quantitative tests focus on evaluating each individual instrument's ability to detect *cocaine and heroin*. An instrument's effectiveness is directly related to the efficiency of the sample collection design; the NDTA tests therefore require a Manufacturer's Representative (MR) to perform sample collection, as well as operate the instrument during the tests. Small quantities of narcotic are dispensed onto test surfaces from which the MR collects samples.

Present technologies only allow for the *convenient* weighing of solids down to a tenth of a milligram. A neat narcotic cannot be accurately dispensed in lesser quantities by weighing the narcotic directly. Alternative techniques are required to allow for the precise conveyance of nanogram to microgram quantities of narcotic. Narcotic test standards must satisfy three criteria:

- The standard must enable the accurate and repeatable dispensing of cocaine and heroin in quantities ranging from subnanograms to micrograms.
- The standard must be reproducible.
- The standard itself should not affect the test results.

The standard should not suppress the performance of a particular vendor's instrument due to chemical or physical properties that are inherent to the standard and not to the narcotic. Equally important is the property of reproducibility. Narcotic standards prepared on a particular day need to behave identically to narcotic standards prepared the same way, from the same starting materials, on any other day.

Narcotic standards that have been explored are categorized as follows:

- Solution-deposited substrate standards,
- Solution-deposited substrate-free standards,
- Vapor-deposited standards,
- Suspension standards, and
- Dry mix standards.

Solution-deposited substrate standards are narcotics coated onto a particulate substrate material while dissolved in a solvent, after which the solvent is evaporated. The narcotic standard is then transferred to the target surface as a solid powder. Solution-deposited substrate-free standards are narcotics deposited onto the target surface while dissolved in a solvent, after which the solvent is evaporated. The narcotic standard is transferred to the target surface as a liquid. Vapor-deposited standards are narcotics deposited by condensation from the vapor phase onto the target surface. Suspension standards are insoluble narcotics deposited onto the target surface while suspended in a liquid, after which the liquid is evaporated. Dry mix standards are solid phase narcotics homogeneously mixed with a particulate filler material; the standard is transferred to the target surface as a solid powder. Experimental characterization of each of these narcotic standards is described in the following sections.

2. SOLUTION-DEPOSITED SUBSTRATE STANDARDS

Solution-deposited substrate standards were originally developed by Revenue Canada[5] and utilized during the first two NDTA test rounds[1,2]. Preparation of these standards involves dissolving the narcotic in a solvent, adding an insoluble solid substrate, and then boiling away the solvent in a manner which homogeneously distributes the narcotic across the substrate surface. A conventional laboratory analytical balance is then utilized to weigh 10 milligram batches of the resulting standard; the amount of narcotic contained within a 10 milligram sample may be varied as desired (typically from 1 nanogram to 10 micrograms). Advantages and disadvantages of solution-deposited substrate standards are outlined in Table 1.

Table 1. Advantages and Disadvantages of Solution-Deposited Substrate Standards

Advantages	Disadvantages
Easy to prepare and to dispense	Poor reproducibility with some substrates
Acceptable longevity at high narcotic mass loadings	Poor longevity at lower narcotic mass loadings
High degree of homogeneity	Enhanced temperature dependence of thermal desorption
Easily wipe sampled and vacuum sampled	A secondary substance is present

Diminished narcotic sensor response may be encountered with solution-deposited substrate standards due to the chemical and physical interactions of the narcotic with the substrate. The magnitude of these effects vary depending upon the substrate chosen. Substrates which have been tested include the following:

- Various sands and particle sizes,
- Fullerenes (Bucky Balls),
- Copper powder,
- Nickel powder,
- Silanized glass beads,
- Monel powder,
- Cellulose powder,
- Silver-coated nickel powder, and
- Pulverized filter paper.

Each of these substrates are treated separately in the following sections. The hydrochloride salts of each narcotic were utilized in these studies.

2.1. Sand substrates

Sand substrates were prepared by placing 20 grams of sand in a tungsten carbide holder with two tungsten carbide balls and pulverizing the material on a Spex Mill for 10 minutes. The crushed sand was then wet-sieved in series through 100μm, 63μm, 45μm and 20μm sieves using acetone. All sand fractions were washed with hexane, chloroform, and methanol to remove organic impurities. Equal amounts of sand retained by the 63μm, 45μm, and 20μm sieves were mixed together for use with cocaine. Sand that passed through the 20μm sieve was used as the substrate for heroin. A narcotic standard was prepared by placing one gram of sand in a flask along with 10mL of a narcotic/methanol solution, after which the mixture was rotovapped to dryness. The concentration of the narcotic/methanol solution was prepared so that a 10mg sample of the dried sand would contain a known mass of narcotic (*i.e., narcotic mass loading*).

Advantages of sand substrates are that they are readily obtainable, inexpensive, and identifiable as a *real-world* substrate. Sand substrates are also straightforward to prepare using common laboratory techniques and equipment. However, one disadvantage of sand substrates is repeatability. Sand is composed of porous particles of silicon oxide. Contaminant molecules settle in the pores and/or attach to the oxygen atoms of the silicon oxide matrix. The exact chemical content of sand is, therefore, dependent upon where the sand was obtained and how the sand was processed. Specifying one of the purist sands, *Ottawa sand,* does not guarantee sufficient consistency. Commercial processing of *washed and dried sand* involves repeated crushing of the sandstone while continually washing with water. Afterwards, the sand is heated to remove the water. Different processing plants use different washing and crushing techniques. There is no specified drying temperature or length of drying time. Flocculants and coagulants may or may not be added to the recycling wash water for environmental reasons. A sand product obtained from a single source is not guaranteed to be identical with each purchase.

Ignited sand is sand that has been heated to extreme temperatures to remove the majority of bound material. Ignited sand, therefore, has a greater number of unbound and loosely bound active sites which interact with the narcotic strongly enough that *zero* signal is obtained from narcotic sensors employing thermal desorption. Figure 1 shows IMS signals as a function of desorber temperature for several substrates. In Figure 1, zero signal was obtained from cocaine/ignited sand standards when analyzed with a commercial ion mobility spectrometer (IMS) system at desorption temperatures up to 284°C. *Washed sand* at 100ng cocaine mass loadings also produced zero IMS signals at desorption temperatures up to 284°C *when simply placed on top of a filter and analyzed.* When wipe-sampled, *washed sand* produced positive cocaine signals at 100ng cocaine mass loadings. This result indicates that wipe sampling abrades narcotic from the surface of these types of standards.

Figure 1. IMS Amplitude vs. Desorber Temperature for a Variety of Solution-Deposited Substrate Standards. (100ng Cocaine Mass Loading). All standards were placed on top of a wipe filter for analysis.

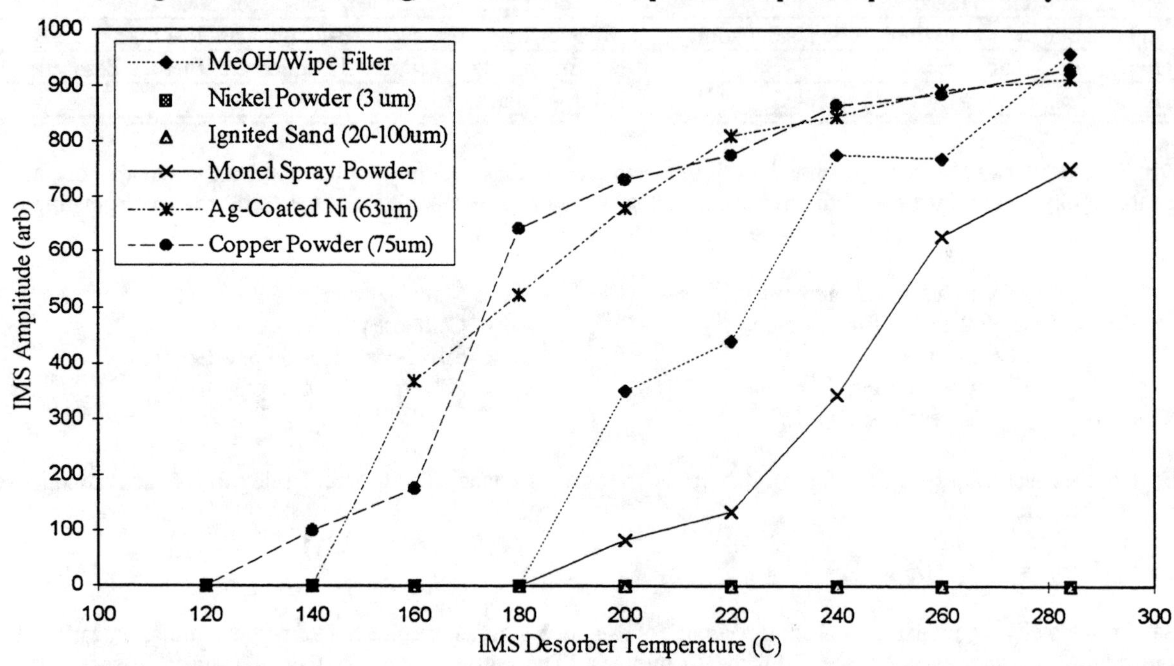

Table 2 illustrates IMS results following wipe sampling of standards produced from two different bottles of *washed and dried* sand having an identical lot number. Bottle #1 had been routinely opened and closed for approximately one year in a laboratory environment. Bottle #2 had remained unopened on a shelf in the same laboratory for the same time period. Heroin standards produced using sand from bottle #2 resulted in markedly lower IMS signals than heroin standards produced using sand from bottle #1. For these reasons, sand standards used for each of the first two NDTA test rounds were prepared from sand originating from a single source.

Table 2. IMS Response for Heroin/Sand Solution-Deposited Substrate Standards from Two Different Bottles of Sand. (284°C Desorber Temperature). Standards were placed on top of a wipe filter for analysis.

Sand Bottle Number:	Bottle #1	Bottle #2
Sand Lot Number:	76843	76843
IMS apparent limit of detection for heroin, following wipe sampling of heroin/sand standards:	500ng	1000ng

Sand substrates are difficult to sieve reproducibly. Each sieve has a theoretical particle size above which material will be retained. Sand was ground with a mortar and pestle, sieved, and then inspected using optical microscopy. A large proportion of smaller sand particles were observed to be retained by each sieve, possibly due to electrostatic clumping. The average particle size retained was substantially smaller than the mesh size of the sieve. The same phenomena was observed with wet-sieving, dry-sieving, and ultrasonic sieving. The presence of these smaller particles results in a greater surface area per narcotic loading than expected, as detailed in Table 3.

Table 3. Surface Area per 10mg of Various Substrate Materials

Material	Density (gms/cm^3)	Average Particle Diameter (μm)	Calculated Number of Particles in a 10mg Batch	Total Surface Area per 10mg Batch (cm^2)
Sieved Ottawa Sand, 63μm - 100μm Theoretical Diameter	1.4	81.5	2.5 x 10^4	5.2
Sieved Ottawa Sand, 63μm - 100μm Observed Diameter	1.4	8.3	2.4 x 10^7	501.4
Powdered Filter Paper	0.44	50	3.5 x 10^5	27.5
Silver-Coated Nickel	9.14	63	8.4 x 10^3	1.0

Sand substrates exhibit shorter longevity at lesser narcotic loadings. Longevity data of cocaine and heroin standards utilizing *washed and dried* sand substrates is illustrated in Figure 2. Decreased longevity at lesser narcotic mass loadings may be due to surface interactions of the narcotic with the substrate.

Figure 2. Longevity of Narcotic/Sand Solution-Deposited Substrate Standards. Analysis performed by gas chromatography / mass spectrometry.

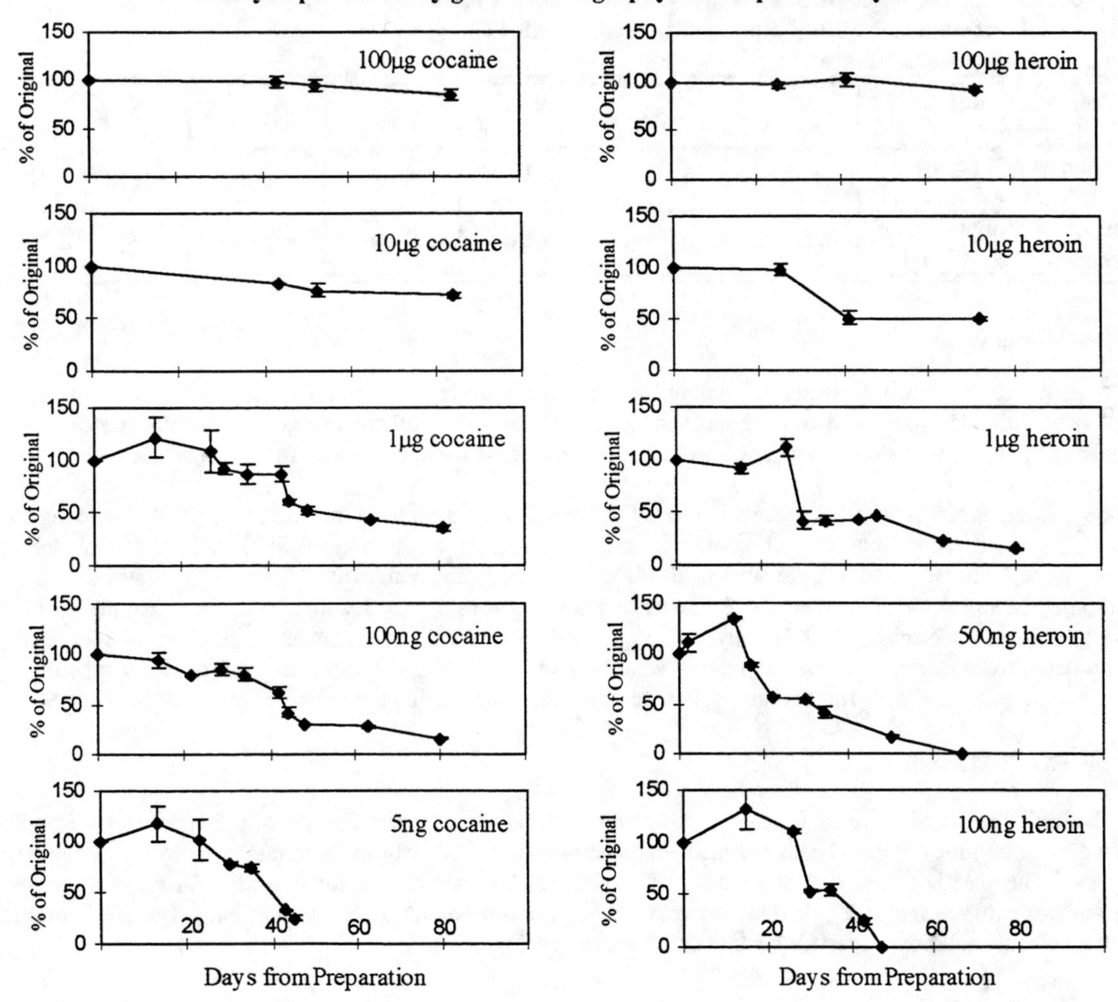

Table 4 presents a simplified approximation of the number of cocaine monolayers per average sand particle at various cocaine mass loadings. The estimation assumes an average rectangular area of a cocaine molecule of 9 angstroms by 15 angstroms and calculates the number of cocaine molecules required to cover the surface of a spherical sand particle. One monolayer of cocaine is achieved when side-by-side cocaine molecules exactly cover the sand particle a single time. The first layer of cocaine molecules experience the maximum physical and chemical interaction with the substrate. Additional layers experience diminishing interaction with the substrate. The calculation predicts that a single monolayer will be achieved at a cocaine loading of approximately 1800ng (1.8μg) per 10mg of 83μm diameter sand. This value correlates well with the pronounced decrease in longevity observed at cocaine loadings of 1μg or less, apparent in Figure 2. Two cocaine loadings in Figure 2 are greater than 1.8μg. These loadings have markedly longer longevity, and correspond to 5.8 monolayers (10μg loading) and 58 monolayers (100μg loading).

Table 4. Cocaine/Sand Solution-Deposited Substrate Monolayer Calculation

Average particle diameter:	8.3×10^{-4} cm
Average volume of 1 sand particle:	3.0×10^{-10} cm^3
Density of Sand:	1.5 gms/cm^3
Average weight of 1 sand particle:	4.5×10^{-10} gms
Average number of sand particles in a 10mg standard:	2.2×10^{7} particles
Average surface area of 1 sand particle:	2.2×10^{-6} cm^2
Average total surface area in a 10mg sand standard:	48.4 cm^2
Estimated cocaine molecular area:	1.4×10^{-14} cm^2

Cocaine Loading per Standard (ng)	1	5	100	500	1000	10000	100000
Number of Cocaine Molecules per Standard	$2x10^{12}$	$1x10^{13}$	$2x10^{14}$	$1x10^{15}$	$2x10^{15}$	$2x10^{16}$	$2x10^{17}$
Number of Monolayers of Cocaine per Sand Particle	**0.00058**	**0.0028**	**0.058**	**0.28**	**0.58**	**5.8**	**58**

2.2. Fullerenes

Fullerenes are a class of molecules including the recently discovered soccer-ball shaped C_{60}, commonly referred to as *bucky ball.* Fullerenes possess a fully conjugated outer shell and are difficult to react, suggesting possible inertness if utilized as a substrate for narcotic standards. In bulk, fullerenes have the appearance of soil, or fireplace soot.

Solution-deposited substrate standards utilizing fullerene substrates were prepared by gently breaking clumps of purified C_{60} (SES, Incorporated) with a mortar and pestle, prior to mixing with narcotic/methanol solutions and rotovapping to dryness. Standards prepared with fullerenes exhibit short longevity. Figure 3 illustrates the longevity of a heroin/fullerene solution-deposited substrate standard with a 1μg heroin mass loading. The mass loading of this standard decays to 50% of its original value in only ten days. At 20ng narcotic mass loadings, *zero* signal was obtained from solution-deposited fullerene substrate standards when analyzed by an IMS employing thermal desorption. Figure 4 illustrates zero cocaine signals from cocaine/fullerene standards at desorption temperatures up to 284°C.

2.3. Silanized glass beads

Silanized glass beads (Sigma Chemical Company) were explored as solution-deposited substrates. Standards were prepared by combining the glass beads (80μm average particle diameter) with narcotic/methanol solutions and rotovapping to dryness. The glass bead standards were analyzed by placing 10mg samples on top of a wipe filter and analyzing by IMS. The standards exhibited enhanced thermal desorption properties relative to sand substrates, but performed poorly relative to narcotic/methanol solutions deposited directly onto filter paper (Figure 4).

Figure 3. Longevity of Heroin/Fullerene Solution-Deposited Substrate Standards. Analysis performed by gas chromatography / mass spectrometry.

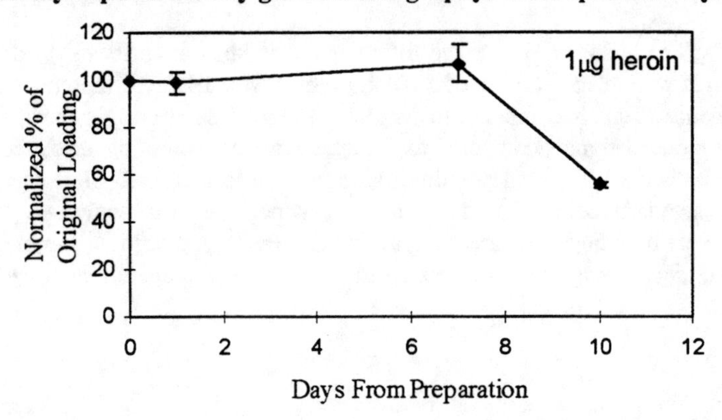

Figure 4. IMS Amplitude vs. Desorber Temperature for a Variety of Cocaine Solution-Deposited Substrate Standards. (20ng Cocaine Mass Loading). Standards were placed on top of a wipe filter for analysis.

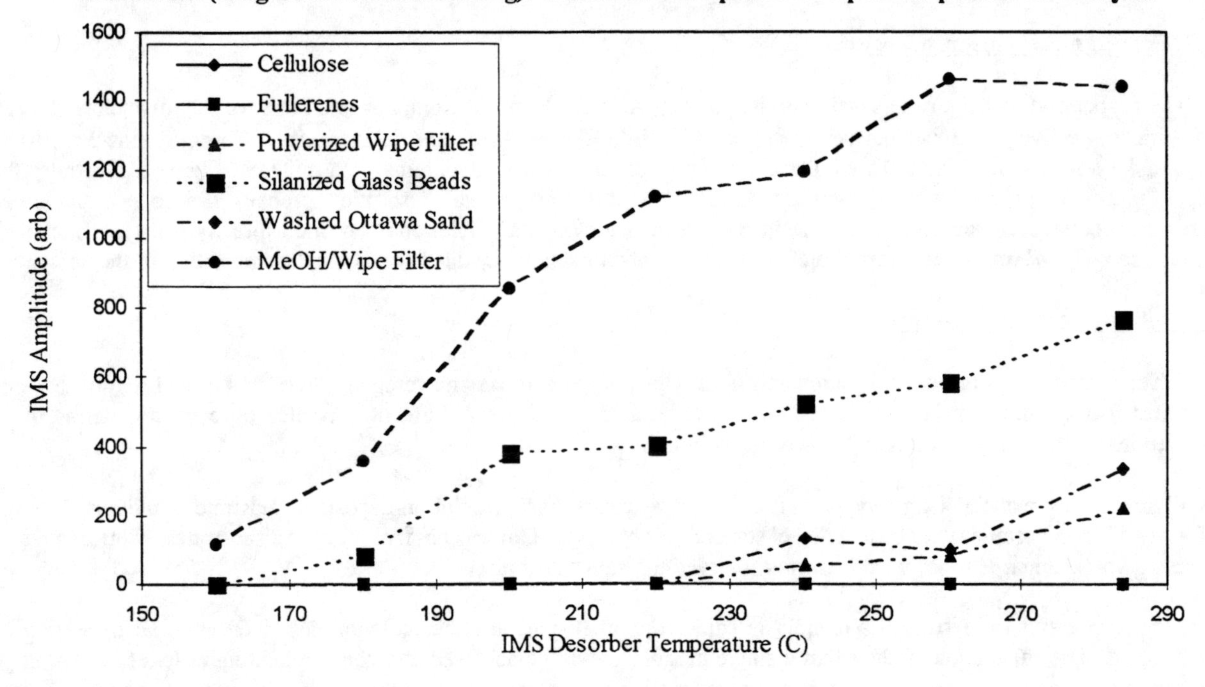

2.4. Cellulose powder

Cellulose powder (Sigma Chemical Company) was combined with narcotic/methanol solutions and rotovapped to dryness. The cellulose powder had an average particle diameter of 50µm. The resulting standards had a pulp-like consistency, and produced small amplitude signals at higher desorber temperatures when analyzed by IMS (Figure 4).

2.5. Pulverized filter paper

Filter paper commonly utilized for wipe sampling by IMS systems (Schlichter & Schuell) was pulverized and utilized as a substrate for solution-deposited substrate standards. A single filter paper was cut into $0.3cm^2$ pieces and placed into a stainless steel canister in 100mg batches. A carbide ball was placed in the canister and the filter paper pulverized on a micro-wiggler mill for five minutes. The resulting powdered filter paper was placed in a flask with narcotic/methanol solutions and rotovapped to dryness.

Figure 4 illustrates the IMS results from pulverized filter paper standards at various IMS desorber temperatures. 10mg of the cocaine loaded standards were placed on top of a second virgin wipe filter for these analyses. Powdered filter paper narcotic standards exhibit reduced signals compared to standards utilizing sand or glass bead substrates. Pulverized filter paper substrates also exhibit reduced signals compared to narcotic/methanol solutions deposited directly on a filter wipe of the same composition. The reduced IMS signal amplitudes are not due to increased filter paper mass, which occurs from the pulverized filter paper standard being placed on another wipe filter for analysis. This was confirmed when narcotic/methanol solutions were deposited onto a wipe filter, which was then placed on top of a second wipe filter prior to IMS analysis. The same IMS signal magnitude was observed with two filter papers that was observed with a single filter paper.

2.6. Metal powders

Solution-deposited substrate standards utilizing metal powders were prepared by combining the metal powder with narcotic/methanol solutions, and rotovapping to dryness. Standards were prepared with copper, nickel, monel, and silver-coated nickel powder substrates.

2.6.1. Copper, nickel, and monel powders

The copper and nickel powders were purchased from Aldrich Chemical Company and have particle diameters of 75μm and 3μm, respectively. Monel powder was purchased from Alpha Aesar and is marketed as a spray powder. Metal standards (100ng cocaine loading) were placed on top of paper wipe filters and analyzed by IMS. Figure 1 illustrates that copper, nickel and monel substrate standards each have different thermal desorption properties. Copper and monel substrates produced greater signals on the IMS than methanol/narcotic solutions deposited directly on wipe filters. In contrast, *zero signal* was observed from nickel substrate standards, possibly due to catalytic decomposition of the narcotic.

2.6.2. Silver-coated nickel powder

Silver-coated nickel powder was purchased from Alpha Aesar and has an average particle diameter of 63μm. Figure 1 illustrates that cocaine standards prepared with silver-coated nickel powder are only rivaled by copper in terms of the observed IMS signal as a function of desorber temperature.

Figure 5 displays the longevity properties of silver-coated nickel solution-deposited substrate standards. Longer longevity of these cocaine silver-coated nickel standards is observed than was observed for sand substrates; shorter longevity of these heroin standards is observed than was observed for sand substrates.

Table 5 presents the results of a monolayer calculation analogous to the calculation described for sand in Section 2.1 and Table 4. The calculation predicts that a single monolayer will be achieved at a cocaine loading of less than 100ng per 10mg of silver-coated nickel. This value correlates well with the pronounced increase in longevity observed at cocaine loadings of 100ng and greater, apparent in Figure 5. Decreased longevity results when cocaine mass loadings produce less than a monolayer.

Table 5. Cocaine/Silver-Coated Nickel Solution-Deposited Substrate Monolayer Calculation

Cocaine Loading per Standard (ng)	1	5	100	500	1000	10000	100000
Number of Monolayers of Cocaine Silver-Coated Nickel Particle	**0.028**	**0.14**	**2.8**	**14**	**28**	**280**	**2800**

Figure 5. Longevity of Narcotic / Silver-Coated Nickel Solution-Deposited Substrate Standards. Analysis performed by Gas Chromatography / Mass Spectrometry.

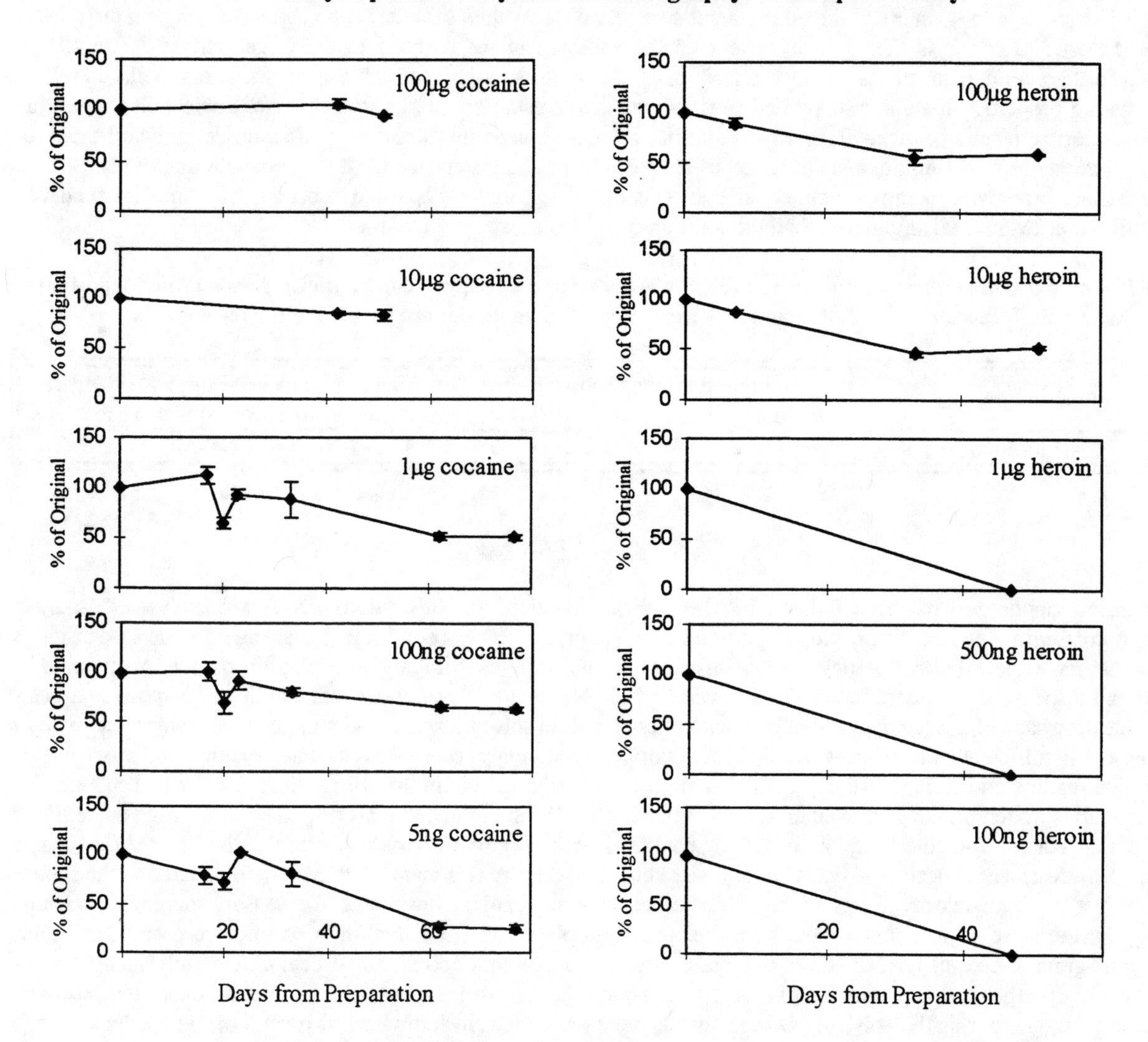

3. SOLUTION-DEPOSITED SUBSTRATE-FREE STANDARDS

Solution-deposited substrate-free standards are produced by dissolving the narcotic in a volatile organic solvent at a known concentration. The narcotic standard is dispensed by transferring a specified volume of the solution (typically with a syringe) onto a target surface. The solvent evaporates leaving the solid narcotic behind. Standards prepared in this manner are functionally identical to the solution-deposited substrate standards previously described in Section 2; in place of a powdered solid substrate, the single target surface acts as a solid substrate. Table 6 outlines the advantages and disadvantages of this type of standard.

Table 6. Advantages and disadvantages of solution-deposited substrate-free standards

Advantages	Disadvantages
Easy to prepare and to dispense	High surface adhesion; very difficult to vacuum sample
High homogeneity and high reproducibility	Adsorption properties dependent on target surface

A sensor that employs vacuum sampling may be incapable of gathering the narcotic for analysis from a solution-deposited substrate-free standard. While solution-deposited substrate particle standards may be vacuum sampled by a sensor, it is not possible to "gather up" the entire target surface with a vacuum sampler. The ability to sample a target with a solution-deposited narcotic is therefore dependent on the adhesion of the narcotic to the target surface. The NDTA evaluations utilize aluminum plates as target surfaces. Solution-deposited substrate-free standards will, therefore, approximate the properties of solution-deposited substrate standards using an aluminum substrate. Table 7 illustrates this point by comparing results obtained from wipe-sampling a solution-deposited cocaine hydrochloride standard from an aluminum plate, with wipe-sampling a solution-deposited particle substrate standard from the same aluminum plate. Both wipe-samples were analyzed using a commercial IMS system. The solution-deposited substrate-free standard produced substantially lower IMS signal amplitudes than the solution-deposited substrate standard.

Table 7. Comparison of a solution-deposited substrate standard using silver-coated nickel to a solution-deposited substrate-free standard. Both standards were placed on an aluminum surface and wipe sampled.

Type of narcotic standard	Solution-deposited substrate-free standard using a cocaine/MeOH soln.	Solution-deposited substrate standard using a silver-coated nickel substrate
IMS signal amplitude (arb)	98	800

5. SUSPENSION STANDARDS

Suspension standards have been utilized by the Federal Aviation Administration (FAA) for explosives[8] and are prepared by combining the insoluble explosive particles with a liquid to create a solid/liquid suspension. The suspension is stirred to distribute the explosive particles in the suspension (subject to centrifugal forces), and a pipette is employed to withdraw and dispense the standard onto a target surface. The volatile liquid evaporates leaving the dry explosive particles behind. The drawback of this technique applied to narcotics is that no solvent has been identified in which cocaine (base or hydrochloride) is sufficiently insoluble. Cocaine hydrochloride is commonly reported as insoluble in ether and oils. Oils do not evaporate readily, and cocaine hydrochloride was determined to be soluble in anhydrous diethyl ether at 1.5ng per 1.0 microliter at ambient temperature. The ether must be handled in a dry environment at all times due to the high solubility of water vapor in ether. The ether may be artificially saturated with cocaine hydrochloride and chilled to reduce cocaine solubility. However, when cocaine particles are deposited onto a surface in this manner, they remain adhered to the surface when subjected to high-velocity blasts of air. Therefore, this standard is unsuitable for sensors employing vacuum sampling. Furthermore, if the ether evaporates in the open atmosphere, water condensation occurs, and cocaine is soluble in water (one gram of cocaine base dissolves in 600mL of water; one gram of cocaine hydrochloride dissolves in 0.4mL of water). Unless the ether is evaporated in a dry inert atmosphere, the standard is identical to a solution-deposited substrate-free standard using water as the solvent. Other solvents were attempted, including tetrahydrofuran (THF), hexane, and other alkanes. Cocaine hydrochloride was found to be soluble in hexane at ~10ng/μl, and greater solubility was observed in THF.

4. VAPOR-DEPOSITED STANDARDS

Vapor-deposited standards are prepared by pulsing known quantities of cocaine vapor from a vapor generator onto a substrate. The substrate can either be a target surface, or a particulate sample. To investigate this technique, an Idaho National Engineering Laboratory (INEL) vapor generator (VG)[7] was loaded with cocaine base. This INEL VG was repeatedly calibrated until the output from the VG was constant from day to day. Calibrations were performed by pulsing the VG for five seconds at a set temperature and flow rate combination onto Teflon sample filters. The Teflon sample filters were then analyzed with a commercial IMS that had been precalibrated using cocaine/methanol solutions.

A mathematical model was derived which closely matches the measured output of the VG. The model is based on the Ideal Gas Law, which can be expressed in terms of moles of narcotic,

$$n_{narc} = P_{narc}V/RT, \quad (1)$$

where P_{narc} ≡ vapor pressure of narcotic, V ≡ volume, n_{narc} ≡ moles of narcotic, R ≡ the ideal gas constant, and T≡ temperature. The mass of the narcotic is obtained by,

$$gms_{narc} = MW_{narc} n_{narc} \quad (2)$$

where gms_{narc} ≡ grams of narcotic, and MW_{narc} ≡ the molecular weight of narcotic. Volume can be expressed in terms of the VG flow rate and pulse length,

$$V = (\text{flow rate})(\text{pulse time}). \quad (3)$$

Substituting equations (2) and (3) into equation (1), the VG output in terms of narcotic mass becomes

$$gms_{narc} = (\text{flow rate})(\text{pulse time})(MW_{narc})P_{narc}/RT. \quad (4)$$

Antoine's equation is commonly used to approximate the vapor pressure of a solid,

$$\log P_{vap} = A - B/(T + C), \quad (5)$$

where A, B, and C are experimentally derived constants, and T is temperature. Rearranging equation (5) and substituting for P_{narc} in equation (4), a generalized equation for the mass quantity of narcotic in a vapor pulse from the vapor generator is obtained:

$$gms_{narc} = (\text{flow rate})(\text{pulse time})(MW_{narc})10^{A}10^{[-B/(C+T)]}/RT. \quad (6)$$

Inserting the gas constant and accounting for units, an equation for the VG output in picograms of narcotic (pg_{narc}) as a function of temperature, flow rate, and pulse length is obtained,

$$pg_{narc} = [(2.672\text{x}10^{5})10^{A}(\text{flow(cc/min)})(\text{pulse(sec)})(MW_{narc}(\text{gms/mole}))10^{[-B/(C+T(K))]}]/T(K) \quad (7)$$

Values for the A and B constants were obtained from the work of Lawrence, *et.al.*[6], which for cocaine are 13.02 and 5884, respectively. The C constant was varied empirically to match the data, resulting in an optimal value of -2.30.

4.1. Vapor-deposited standards condensed onto a surface

The INEL VG was used to grow cocaine particles on a glass surface. The VG nozzle was directed vertically downward onto a silanized glass microscope slide and pulsed for varying lengths of time. Equation (7) was used to approximate the VG output. Periodically, the glass microscope slide was viewed with an optical microscope, and particle statistics were recorded from visual observations. As the total pulse time of the VG was increased, the size of the particles increased, indicating that the particles were actually forming and growing on the surface of the microscope slide. Table 8 illustrates the observed particle statistics for cocaine standards generated with this technique.

Vapor-deposited standards condensed onto a glass surface can be easily wiped from the surface, but they remain undisturbed when subjected to energetic blasts of air. The high adhesion of cocaine particles generated in this manner indicates these standards would not be conducive to vacuum sampling. Vapor-deposited cocaine particles were observed to vanish from the silanized glass surface within 24 hours, presumably due to sublimation. Table 9 details the observed disappearance of the vapor-deposited cocaine particles.

Table 8. Vapor-Deposited Cocaine Particle Growth on a Silanized Glass Surface.

Accumulated Pulse Time (seconds)	Deposit Area (mm x mm)	Cocaine Particle Number Density (per 25μm x 25μm)	Cocaine Particle Size Distribution (μm)
100	undetectable	undetectable	undetectable
200	6 x 6	faint	faint
300	6 x 6	9	0.75
400	8 x 8	18	0.75
500	9 x 9	13	1.25
600	10 x 9	14	0.75 - 2.5
700	12 x 10	14	0.75 - 2.5
800	13 x 11	18	0.5 - 2.5
900	13 x 10	16	0.5 - 3.0
1000	13 x 12	20	0.5 - 3.0

Table 9. Sublimation of Vapor-Deposited Cocaine Particles from a Silanized Glass Surface

Time After Completion of Particle Growth (hours)	Deposit Area (mm x mm)	Cocaine Particle Number Density (per 25μm x 25μm)	Cocaine Particle Size Distribution (μm)
0	13 x 12	20	0.5 - 3.0
4.5	8 x 7	19	1.0 - 3.0
19	1 x 1	5	0.75 - 2.5
20	undetectable	undetectable	undetectable

4.2. Vapor-deposited standards condensed onto particulates

The INEL vapor generator was used to deposit cocaine onto particulate substrates. The VG nozzle was directed vertically downward towards a silanized glass microscope slide supporting 10mg of a particulate substrate material. Equation (7) was used to approximate the desired VG narcotic output. Substrate effects similar to those for solution-deposited substrate standards were observed. The following particulate substrates were tested:

- Various sands and particle sizes,
- Fullerenes (Bucky Balls),
- Copper powder,
- Nickel powder,
- Monel powder,
- Cellulose powder,
- Silver-coated nickel powder,
- Pulverized filter paper, and
- Non-pulverized filter paper.

Figure 6a illustrates results obtained for standards generated by vapor-depositing cocaine on a washed sand substrate. These samples were analyzed by placing the standard on a filter wipe and analyzing by IMS. These results are compared to results from vapor-depositing cocaine directly onto a filter wipe. At higher desorption temperatures, the cocaine signal from the sand substrate actually disappears. Cocaine was then vapor-deposited onto a filter wipe, and the same *narcotic-free* sand substrate material was placed *on top of the filter paper* and analyzed by IMS. A completely different signature was obtained as a function of desorption temperature (Figure 6b). This result demonstrates that significant interactions may occur between vapor-deposited cocaine and the substrate onto which it condenses.

Figure 6a. Cocaine Vapor-Deposited Standard on Sand Compared to Cocaine Vapor-Deposited Standard on a Wipe Filter. Samples were analyzed by IMS.

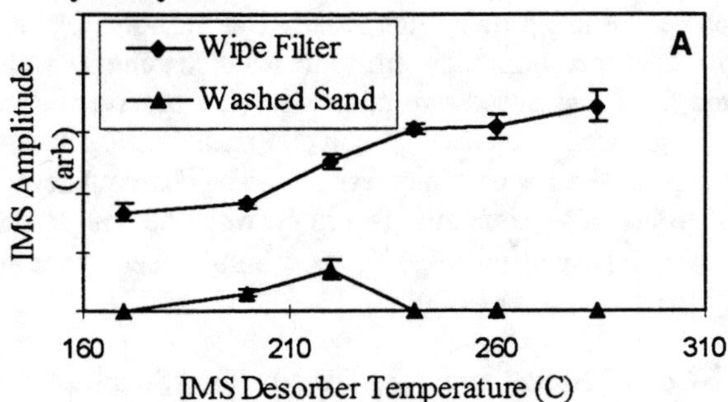

Figure 6b. Cocaine Vapor-Deposited Standard on Wipe Filters. One of the wipe filters was covered with 10mg of sand after vapor deposition. Samples were analyzed by IMS.

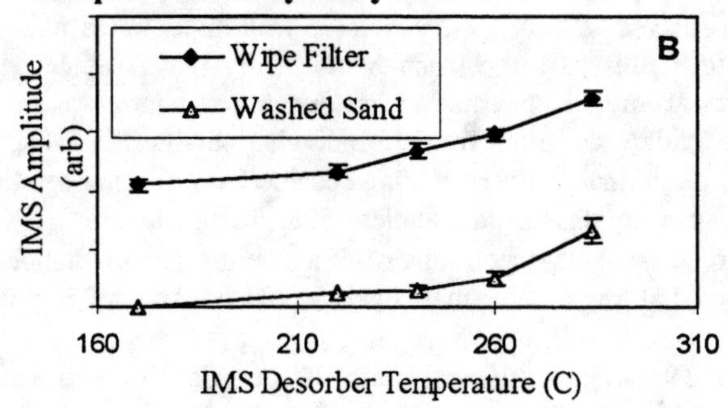

The behavior observed with sand substrates in Figure 6 was explored with other substrate materials. Vapor was deposited onto a filter paper, and then covered with 10mg of narcotic-free substrate material. IMS signals as a function of temperature are shown in Figure 7. Note the maximum IMS signal at 220°C for the silver-coated nickel and monel spray powder substrates. This phenomena was reproduced several times, and occurs at the same temperature as the maxima observed with sand in Figure 6a.

Figure 7. IMS Amplitude vs. Desorption Temperature for Substrates Placed on Top of Vapor-Deposited Cocaine

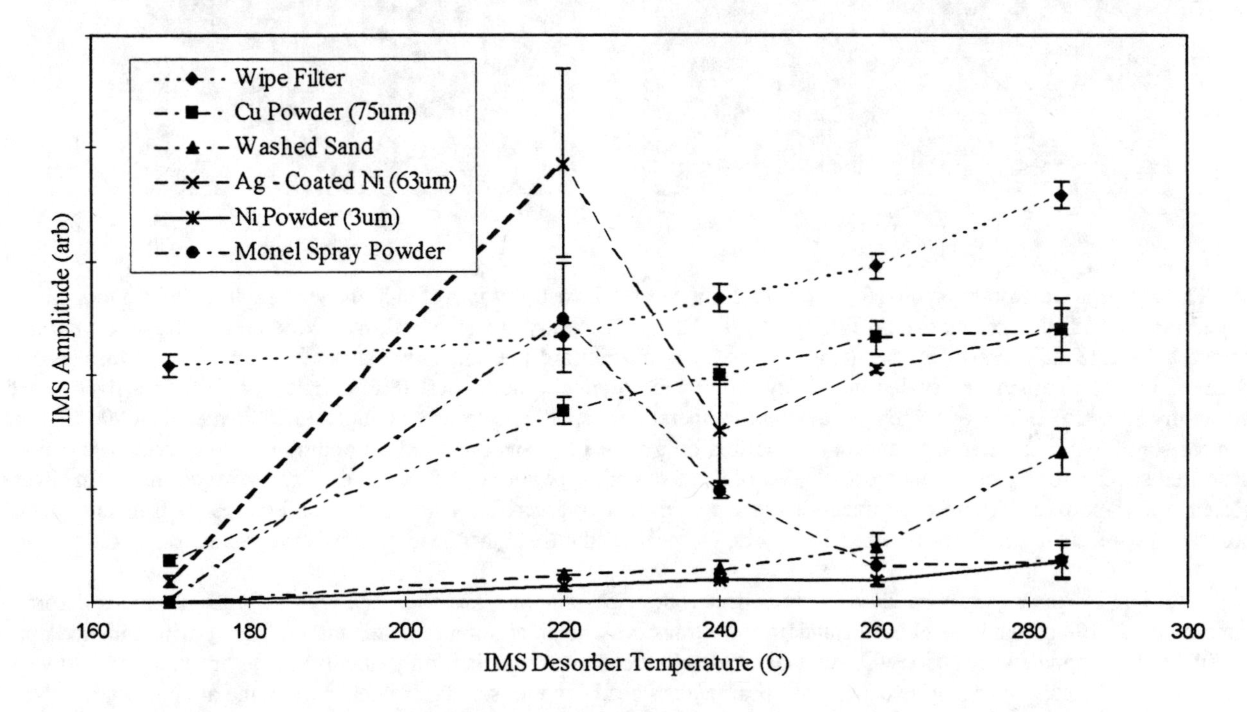

The unusual temperature dependence observed with several of the vapor-deposited cocaine standards was considered problematic. Therefore, further exploration of vapor-deposited standards was abandoned.

6. DRY MIX STANDARDS

Dry mix standards are prepared by combining a known quantity of narcotic with a known quantity of another solid filler material in powder form. The two powders are mixed in a pre-calculated ratio to generate a dry mix standard with a

Following the grinding and sieving of the narcotic, the solid filler powder is added, and the mixture is agitated by a vortex shaker for ten minutes. The mixtures are then homogenized by a micro spinning riffler (Gilson Company, Inc.) ten times. The dry mix batch may be further diluted with filler powder to create successively lesser narcotic mass loadings. Each successive dilution is generated with an identical mixing procedure using the vortex shaker and the spinning riffler. Figure 9 illustrates the batch homogeneity that is achieved with this method at low narcotic mass loadings. A 10ng cocaine/10mg silver-coated nickel dry mix standard was prepared from four successive dilutions. The observed narcotic mass loading deviation from 35 individual analyses is 5.4%. In comparison, the observed narcotic mass loading deviation of 35 individual analyses of 10ng cocaine/1μl methanol solutions deposited on wipe filters was 5%. The observed deviation of 35 heroin/glass bead samples (50ng heroin loading) is 7.6% which also compares favorably with the 3% deviation observed from 35 repetitions of 50ng heroin/1μl methanol samples deposited on wipe filters. Similar deviations were observed at 5ng cocaine mass loadings (8.8%), and at 25ng heroin mass loadings (9.8%).

Figure 9. Dry Mix Homogeneity of 10ng Cocaine/10mg Silver-Coated Nickel and 50ng Heroin/10mg Silanized Glass Bead Standards. Standards were placed on top of a wipe filter and analyzed by IMS.

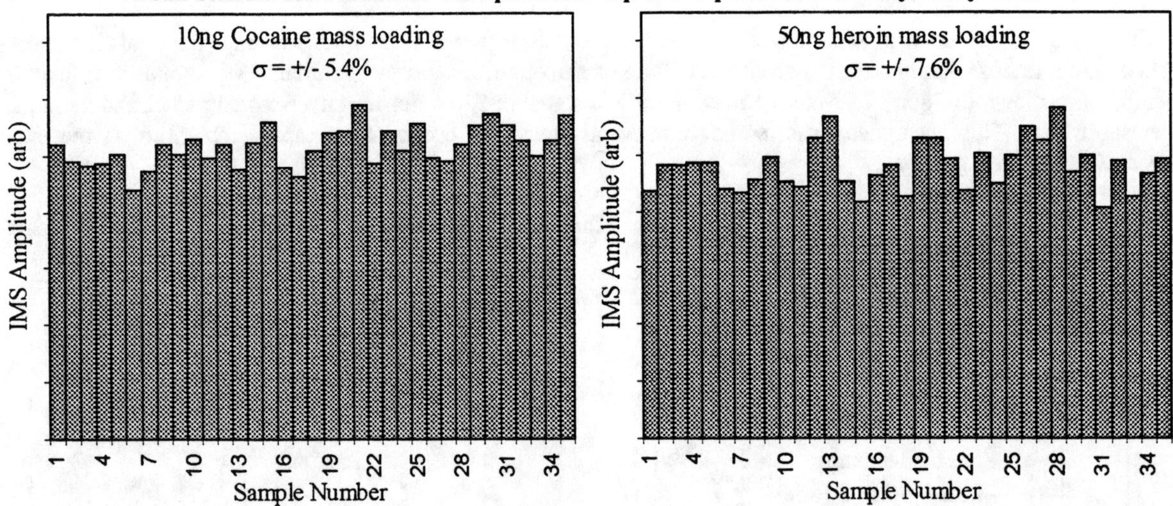

The temperature desorption characteristics of 20ng cocaine loaded dry mix standards were explored with sand, silver-coated nickel, and silanized glass bead filler powders. 10mg of each dry mix standard was placed on top of a wipe filter and analyzed by IMS at various desorber temperatures. The results are presented in Figure 10. The best performance is obtained with a cocaine/silver-coated nickel dry mix, which produces the highest IMS signals. The cocaine/silver-coated nickel dry mix also exhibits a relatively invariant temperature response at desorber temperatures greater than 200°C. For comparison, 20ng cocaine in methanol was deposited on a wipe filter and analyzed, in addition to 20ng cocaine/solution-deposited silver-coated nickel standards. Neither of these standards performed as well as the dry mix of cocaine with silver-coated nickel powder. These experiments were repeated at a 5ng cocaine loading; the differences in IMS amplitudes observed between a cocaine/silver-coated dry mix and a cocaine/silanized glass bead dry mix were more pronounced.

Additionally, the dry mix standards were collected by wipe-sampling and then analyzed by IMS at various desorber temperatures. 10mg quantities of each standard were transferred to an aluminum plate, and repeatedly wipe sampled until no further IMS signals were observed. The magnitude of all observed signals for a particular 10mg dry-mix standard were summed. Figure 11 displays these data. The cocaine/glass bead dry mix standards behave the same as the cocaine/silver-coated nickel dry mix standards. Both of these dry mixes outperform a cocaine/methanol solution deposited directly on a wipe filter, as well as cocaine solution-deposited substrate standards using silver-coated nickel substrates.

Dry mix standards of heroin were explored in the same manner as the cocaine dry mix standards, using 50ng heroin mass loadings per 10mg of powdered filler. Figure 12 presents the results obtained when 10mg standards are placed on top of a wipe filter for analysis. Figure 13 presents the results obtained when the standards are placed on an aluminum target plate and repeatedly wipe sampled. The solution-deposited substrate standards produced either zero signals or very small known narcotic mass loading. Advantages and disadvantages of dry mix standards are outlined in Table 10. The following filler powders were evaluated:

- Various sands and particle sizes,
- Silver-coated nickel powder, and
- Silanized glass beads.

Batch homogeneity at each narcotic mass loading is enhanced by uniform particle sizes of both the narcotic and the solid filler powder. To prepare these standards, the hydrochloride salt of the narcotic is ground with a mortar and pestle, and passed through a 30 micron sieve prior to the addition of the filler powder. Particle size statistics were visually observed through an optical microscope following sieving for both heroin and cocaine. Figure 8 is a photograph of the processed heroin hydrochloride powder. Heroin particles are observed with a nearly uniform diameter of approximately 2μm. Cocaine hydrochloride processing produces particles with three different particle size ranges, centering around diameters of 2μm, 7μm, and 15μm.

Table 10. Advantages and Disadvantages of Dry Mix Standards

Advantages	Disadvantages
Reduced thermal desorption effects	Labor-intensive production
Increased longevity at all mass loadings	A secondary substance is present
Easily wipe sampled and easily vacuum sampled	
Reduced substrate effects	

Figure 8. Optical Microscope Photograph of 2μm Diameter Heroin Particles After Grinding and Sieving. The distance between each scale marker is 2.5μm.

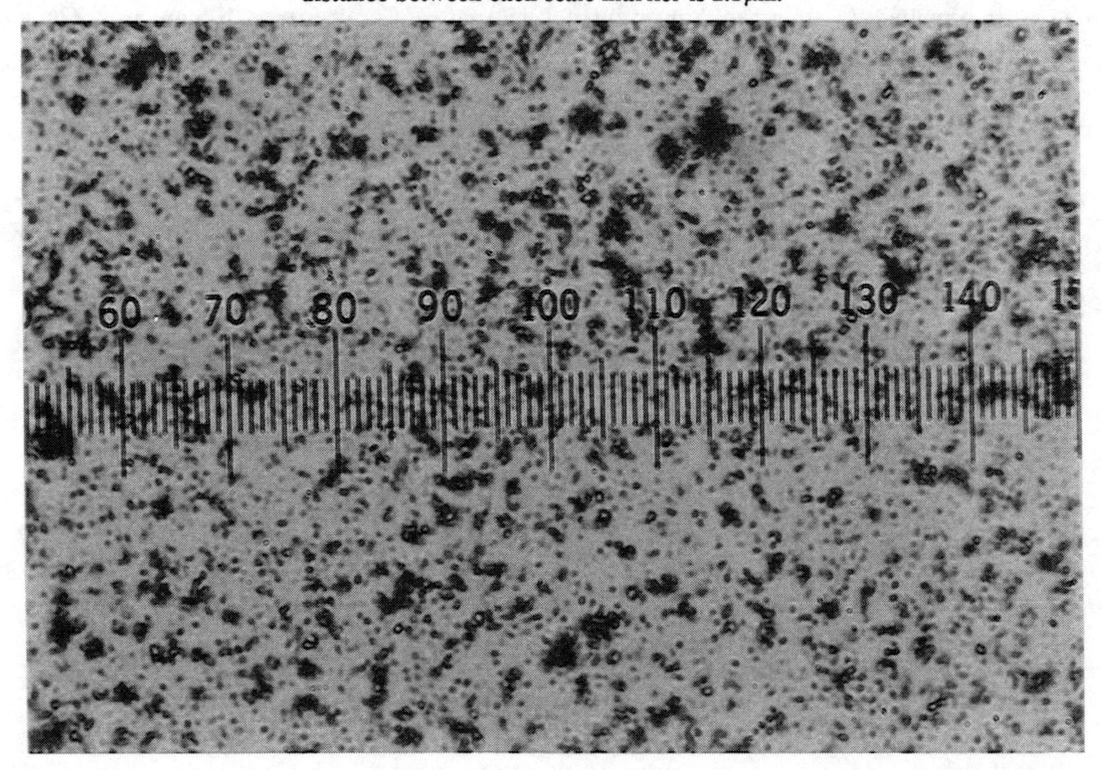

Figure 10. IMS Amplitude vs. Desorber Temperature for a Variety of Cocaine Dry Mix Standards. (20ng Cocaine Mass Loading). All standards were placed on top of a wipe filter for analysis.

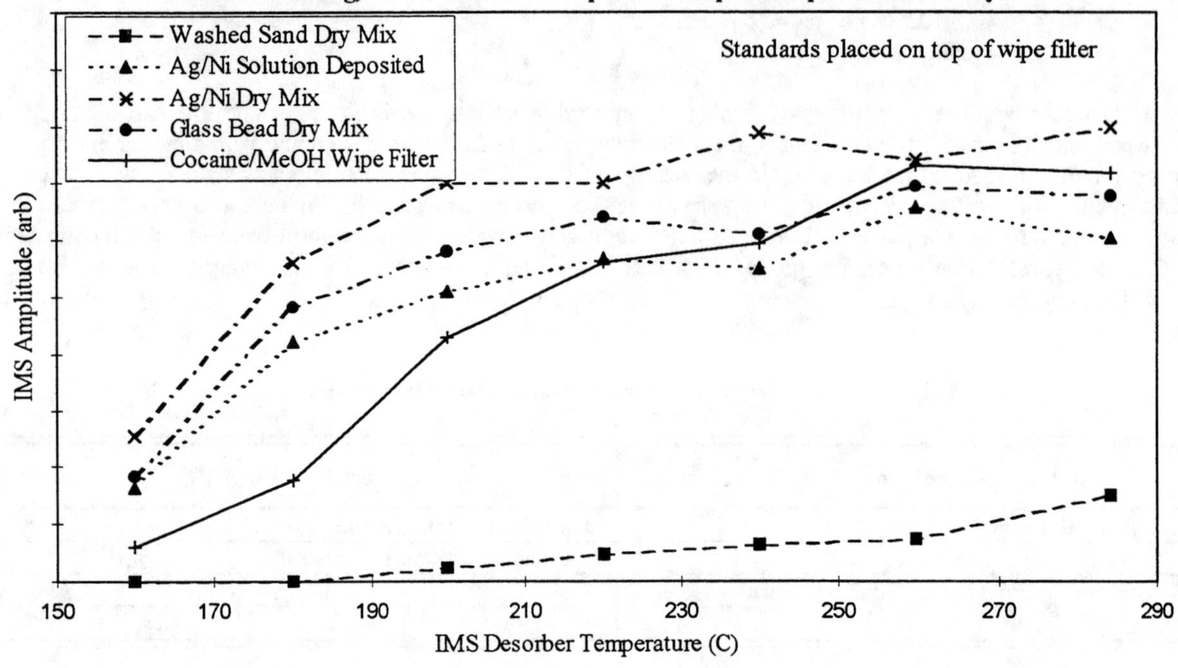

Figure 11. IMS Amplitude vs. Desorber Temperature for a Variety of Cocaine Dry Mix Standards. (20ng Cocaine Mass Loading). All standards were collected for analysis by wipe sampling.

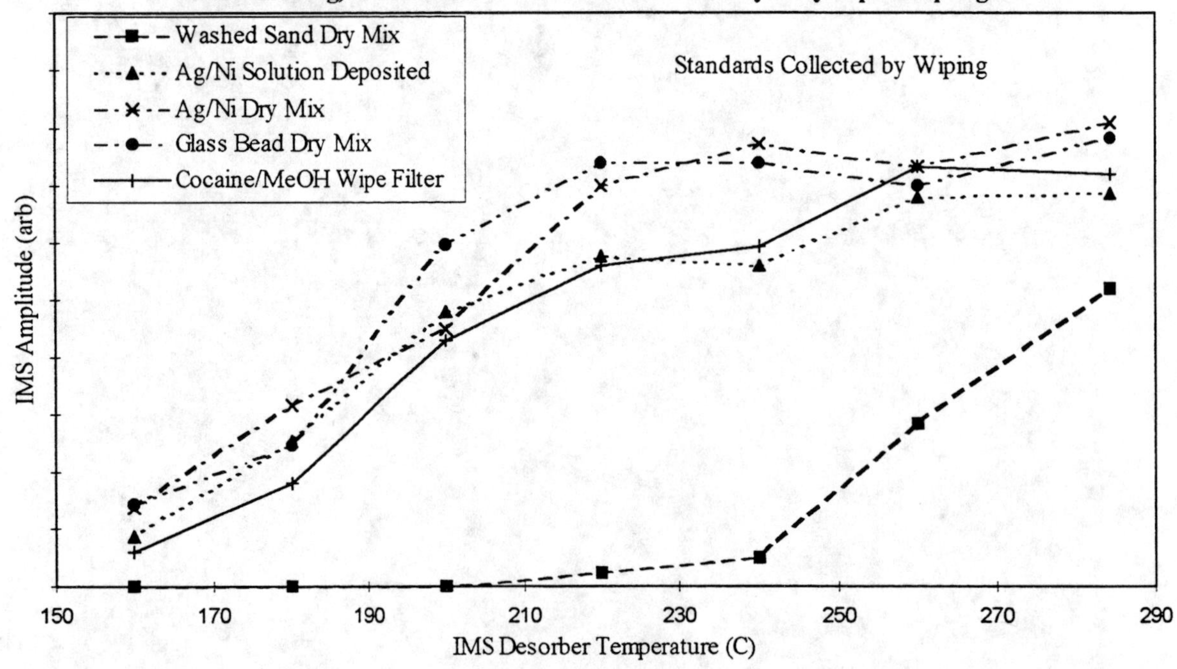

Figure 12. IMS Amplitude vs. Desorber Temperature for a Variety of Heroin Dry Mix Standards. (50ng Heroin Mass Loading). All standards were placed on top of a wipe filter for analysis.

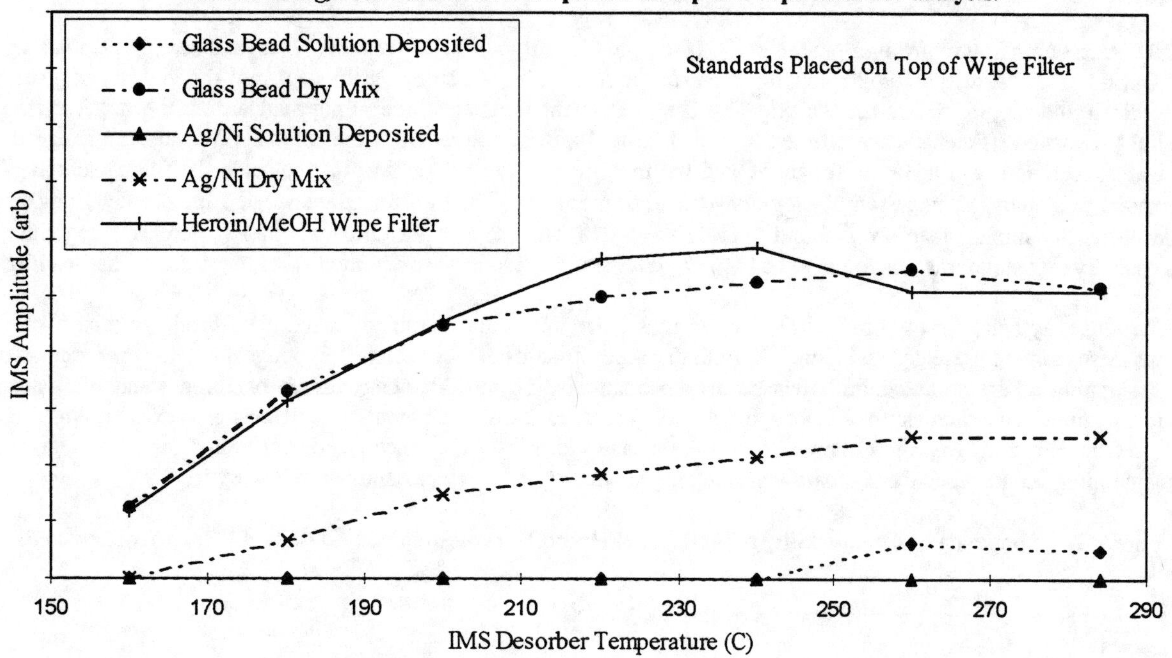

Figure 13. IMS Amplitude vs. Desorber Temperature for a Variety of Heroin Dry Mix Standards. (50ng Heroin Mass Loading). All standards were collected for analysis by wipe sampling.

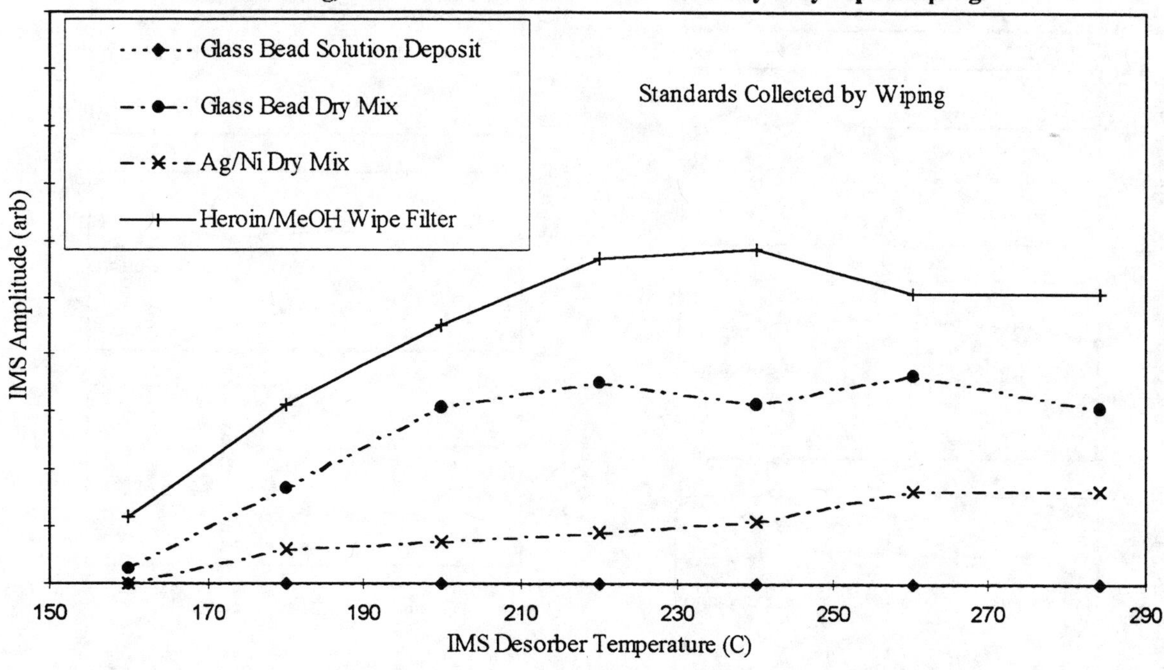

signals for all cases attempted. The best dry-mix performer in both cases is a heroin/silanized glass bead dry mix. A heroin/methanol solution deposited on a filter paper performed as well as the heroin/glass bead dry mix when the dry mix was directly deposited onto a filter paper. The heroin/methanol solution produced higher IMS signals than the silanized glass bead dry mix when wipe sampling was performed, presumably due to transfer efficiency of the heroin from the aluminum target plate to the wipe filter.

NDTA test preparation involves pre-weighing 10mg standards into individual sample vials, and sealing the vials with a cap. During a NDTA test, a 10mg standard is transferred to a target by removing the cap from the vial, turning the vial upside down and tapping it against the target surface. Efficient transfer of the 10mg standard to the target surface is essential. Transfer efficiencies were tested by transferring 10mg standards from the sample vials into centrifuge tubes. Afterwards, methanol was added to the tubes, and the mixture was centrifuged in order to settle the filler material. The supernatant was analyzed by gas chromatography/mass spectrometry. The transfer efficiency of 1μg cocaine/ 10mg silver-coated nickel dry mix standards was found to be 96% +/- 2%, and the transfer efficiency of 1μg heroin/ 10mg silanized glass bead dry mix standards was found to be 94% +/- 2%. Similar results were obtained at lesser narcotic mass loadings.

The longevity behavior of cocaine/silver-coated nickel dry mixes and heroin/silanized glass bead dry mixes was also studied; these data are presented in Figure 14. Both dry mix standards exhibit satisfactory longevity. However, degradation during shipping of dry mix standards has been problematic. Zero to 100% degradation of 10mg standards placed in individual sample vials and shipped via air freight was observed during shipments from Houston, Texas to Washington, DC, and from Houston, Texas to Ottawa, Ontario, Canada. Additional experience has demonstrated that zero degradation during shipping occurs when the standard vials are purged with dry nitrogen gas and packed in dry ice.

Figure 14. Longevity of Cocaine/Silver-Coated Nickel and Heroin/Silanized Glass Bead Dry Mix Standards

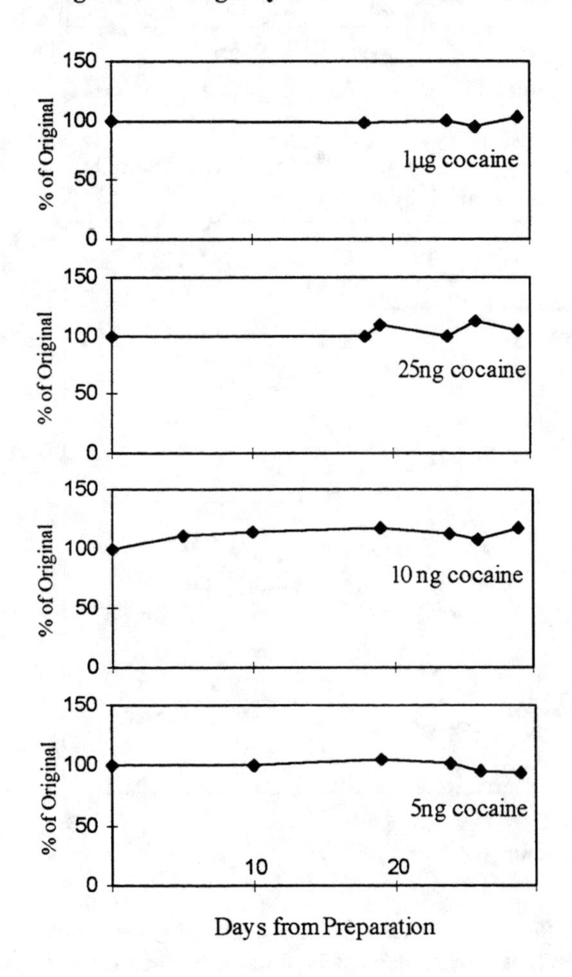

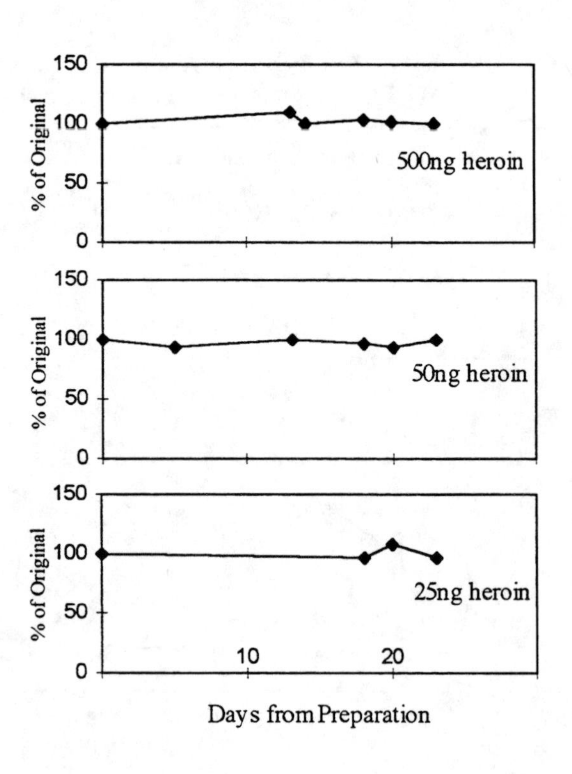

7. CONCLUSIONS

The NDTA program strives to evaluate narcotic detection devices in a manner that is removed from any specific end user application. Test results are interpreted by law enforcement agencies (LEAs) according to the individual LEA's applications. For example, a detection system that is deployed by the US Coast Guard will likely experience a different environment than a system deployed by a correctional facility. This work has demonstrated that substrate effects influence the physical properties of narcotics at the microscopic level. Simultaneously, the exact nature of cocaine contamination in many environments has not been explored; particulate contamination may exist as small independent free particles, hydrated particles, hydrated aerosols, or crystalline films. The physical interaction of each of these moieties with a surface will be different. Therefore, a desirable test standard for the NDTA program is a standard that is inert to its environment. For this reason, the best test standards for heroin and cocaine are of the dry mix type. The optimal standard for cocaine is a dry mix using silver-coated nickel powder. The optimal standard for heroin is a dry mix with silanized glass beads. These standards have a long shelf life, and have been shown to be homogeneous and reproducible.

8. REFERENCES

1. *Benchmark Evaluation Studies of the Illicit Substance Detector, Accupress, Sentor 5000, and Ionscan 350 Drug Detection Devices, November 1995,* Executive Office of the President, Office of National Drug Control Policy, Counterdrug Technology Assessment Center, available ONDCP/CTAC, on a Law Enforcement Sensitive, Official Government Use Only basis.

2. *Benchmark Evaluation Studies of the Barringer Ionscan 400, Graseby Narcotec, Viking SpectraTrak 672, Ion Track Instruments Itemiser, and CPAD Ariel/PID Drug Detection Devices, February 1996,* Executive Office of the President, Office of National Drug Control Policy, Counterdrug Technology Assessment Center, available ONDCP/CTAC, on a Law Enforcement Sensitive, Official Government Use Only basis.

3. *Benchmark Evaluation Studies of the Securetec Drugwipes, ITI Itemiser Contraband Detector, Graseby Narcotec, and Scintrex NDS-2000,* Executive Office of the President, Office of National Drug Control Policy, Counterdrug Technology Assessment Center, available ONDCP/CTAC, on a Law Enforcement Sensitive, Official Government Use Only basis, in press.

4. David Hoglund, C.Su, R.Jadamec, D.Lucero, P.Pilon, P.Neudorfl, G.Drolet, T.Kunz, S.Ulvick, C.Roche, E.Huang, J.Demirgian, S.Rigdon, P.Shier, L.Norwood, "U.S. Customs Service Performance Assessment of Cocaine / Heroin Detection Systems", *Proceedings, Counterdrug Law Enforcement: Applied Technology for Improved Operational Effectiveness International Technology Symposium,* October 24-27, 1995, Nashua, New Hampshire, pp. 1-11 to 1-18.

5. Pierre Pilon, Michel Hupe, Mohinder Chauhan, Andre Lawrence, "Development of Test Material for Narcotics Detection Equipment: Sand/Drug Mixtures," *Journal of Forensic Sciences* **41**(3), pp. 371-375, (May 1996).

6. A.H. Lawrence, L. Elias, and M. Authier-Martin, "Determination of amphetamine, cocaine, and heroin vapour pressures using a dynamic gas blending system and gas chromatographic analysis", *Canadian Journal of Chemistry* **62**(10), p.1986, (1984).

7. J.P. Davies, L.G. Blackwood, S.G. Davies, L.D. Goodrich, and R.A. Larson, "Design and Calibration of Pulsed Vapor Generators for TNT, RDX and PETN", *Advances in Analysis and Detection of Explosives,* pp.513-532, Kluwer Academic Publishers, Netherlands, 1993.

8. Frank T. Fox, Steve Sisk, Richard DiBartolo, Derry A. Green, Jill F. Millar, "Preparation and Characterization of Plastic Explosive Suspensions as Analogs of Fingerprint Derived Contaminants for use in Certification of Explosives Detection Systems", *Proceedings, The Third Workshop of the International Civil Aviation Organization (ICAO) Ad Hoc Group of Specialists on the Detection of Explosives,* October 16-19, 1995, Atlantic City International Airport, New Jersey.

SESSION 2

Detection of Drug Usage

The Keys to a Drug Free Work Place

Joseph J. Fortuna
and
Patricia Boyle Fortuna

Chemical Detection Services, Inc.
7240 F Telegraph Square Drive
Lorton, VA 22079
Phone: 703 550-1806 Fax: 703 550-1808
Email: CDSI@EROLS.COM

ABSTRACT

What does it take to establish a drug free work place? Are technologies available other than urine testing for pre-employment screening and monitoring of employees? Various methods are now available to screen for illicit drug residues on items handled by individuals. The residues can be acquired from the surfaces of items such as telephones, door knobs, steering wheels, lockers, clothing, identification cards, etc. Test kits are also available for urine testing at NIDA threshold levels. Analysis of hair, saliva, and sweat is now possible.

How good are these methods and kits? What value are they to the public? What are the legal concerns facing employers? What do the screening tests show? These questions and others are addressed in this paper.

The authors review for the reader how drug abuse by U. S. workers costs businesses. The paper then addresses the various aspects of the DOT regulations to determine why urine analysis (UA) is insufficient to eliminate drug abuse. The authors present applications of screening technologies in addition to UA. Finally, the authors provide a conclusion of findings and recommendations for businesses that truly want or need drug free work places.

Keywords: Work place drug testing, urine analysis, impairment testing, ion mobilitlity spectrometer, immunassay kits, GC/MS, DOT mandated company,

INTRODUCTION

New technologies, when introduced to law enforcement and ultimately to the general public, need to be evaluated in the user environment. Residue collectors and analyzers, drug screening immunoassay kits, and sprays are now being marketed for law enforcement, customs, and for the general public. The factors influencing the use of these alternative screening tools include awareness, cost benefits, compliancy, fear of litigation, and government guidance. The environments for these markets vary and impact their use.

Law enforcement agencies operate in the domain of "probable cause". A government official, such as a police officer or Federal agent, has observed some behavior or other indication that a crime has been committed. Probable cause is required to obtain a search warrant, which then allows for more thorough searching. The exception to the probable cause protocol in law enforcement is found in the U. S. Customs Service domain. The Customs Service may screen any item or person for contraband at U.S. borders without concern for probable cause.

On the other hand, the work place functions in the realm of "reasonable suspicion." Reasonable suspicion allows a more flexible scenario for the use of technologies. The owner or manager of a business has a greater latitude under the concept of reasonable suspicion than has law enforcement under probable cause. The use of drug screening technologies raises the issues of interpersonal relationships, trust, and individuals' rights of privacy. These need to be addressed in the Drug Policy Statements of a business.

SPIE Vol. 2932 • 0-8194-2334-3/97/$10.00

The major problem facing supervisors and police officers is that they are required to make subjective judgments about employees or citizens based on behavior. Police officers who interview drivers also rely on odors. Alcohol and marijuana are easily detected due to their distinctive odors. Drivers impaired by other illicit drugs are very difficult to identify through behavior and speech.

Businesses that are mandated by the U. S. Department of Transportation (DOT) to have anti-drug programs, such as commercial trucking firms, conduct a drug test when reasonable suspicion is established by the supervisor. Reasonable suspicion determination is based on specific, articulable, contemporaneous, observations concerning the appearance, behavior, speech, or body odors of employees. Again, these are difficult judgements to make, even if trained. There are several methodologies now available to extend or to corroborate reasonable suspicion observations. Confirmation is usually performed through urine analysis.

Urine analysis is the accepted method for the basis of counter-drug programs in the work place. The use of urine analysis dates from 1988 when other technologies were not available. Alternative technologies and sampling targets are available today. The question is, do these newer technologies provide a benefit over urine analysis? This paper evaluates the currently employed urine analysis process and reviews the various alternative or supplemental technologies in the context of the work place. We begin by examining the elements of a drug free work place program.

DRUG-FREE WORK PLACE PROGRAM ELEMENTS

Five elements comprise a drug free work place:

1. Commitment and policy
2. Education and training
3. Deterrence and detection
4. Consequence policies
5. Follow-up procedures

Commitment and policy. Companies that are DOT mandated to have anti-drug programs have a built in commitment and policy guidelines to follow. Companies that voluntarily implement an anti-drug program require a commitment based on internally generated objectives. These objectives could be to increase productivity, decrease accident rates, decrease theft, decrease absenteeism and tardiness, increase co-worker safety, etc. Once the commitment is made, a policy statement is needed. The policy is normally distributed to all new and present employees. In unionized facilities, negotiations between the management and the union are generally required.

Education and training. Educating the work force about drug abuse is usually performed by providing informational materials, community hot line assistance numbers, or the telephone number of an employee assistance program, if available. Employees in safety or security sensitive positions should be trained on the effects and consequences of illicit drugs on health, safety, and about the indicators and symptoms of illicit drug use. Responsible personnel, such as supervisors, should be trained to recognize indicators and symptoms of substance abuse to substantiate and document reasonable suspicion. Reasonable suspicion training includes awareness and identification of physical, behavioral, and performance indicators of drug use.

Deterrence and detection. Urine analysis is the most prevalent method of illicit drug deterrence and detection. In a 1996 survey by the American Management Association [1], 92% of their responders use urine sampling, and 79% use no other method of drug testing. Fifteen percent use blood sampling, and 0.8% use only blood samples. Hair sampling is used by 2% of the responding companies, and 0.5% use only hair analysis. Non-medical performance testing is used by 2% of the companies, and 0.2% use only non-medical performance testing.

Ninety-eight percent of all employees test negative for drugs. More than 90% of these tests are based on urine testing (ODD JOBS, Washington Post, P. H4, Sunday, July 28, 1996). And yet two-thirds of drug users are employed (Drug Strategies, "Keeping Score," p. 20, 1995).

Urine testing programs do deter applicants and most likely deter employees from abusing drugs. Urine analysis is accepted as the standard for anti-drug programs in the corporate world. The U. S. Department of Transportation (DOT) mandates that businesses regulated by DOT components employ UA. The companies needing to comply with the DOT regulations are known as mandated companies. There are also corporations and businesses that use UA because they want an anti-drug program.

The methods, procedures, analytical techniques, reporting, and privacy considerations used by mandated companies are well defined in the Federal Code of Regulations. These regulations serve as useful guides for non mandated businesses. The programs outlined in the Code of Regulations have been used and verified as a system that contributes to the protection of the business, the employee, and the integrity of the safety objective of DOT. But the problem of drug abuse continues in both mandated and non-mandated companies that use UA.

For example, twenty-four percent of full-time employees, when interviewed by the Gallup Organization, reported they had "personally seen or heard of" drug use at work, and 41% stated that drug use in their workplace "seriously affected their ability to get the job done."

Consequences. Failing a drug test during pre-employment screening usually results in the prospective employee being denied employment. Consequences of current employees failing a random or mandated urine test vary among companies. Companies like Black & Decker (Washington Post, p. A1, Sept. 16, 1996) attempt to rehabilitate its employees after a failed test. Other companies dismiss the employee for "cause." Other consequences may involve temporary assignment to non safety jobs. This action is required of mandated employers.

Follow-up procedures. Mandated companies are required to test employees in safety sensitive jobs prior to their returning to duty. Employees, once returned, are then required to be tested periodically over the course of one year.

The American Management Association (Machine Office Technology, July 1996, p.23) reported that drug testing in the work place has risen 277 percent since 1987. The AMA reported [1] that eighty-nine percent of responding manufacturing firms test. In the service sector, 73 percent test. Due to DOT regulations, 100 percent of transportation firms test. Seventy-nine percent of wholesalers and retailers, 60 percent of business service providers, and 56 percent of financial service providers test. However, only 8 percent have conducted cost benefit analyses. Two percent of the responders stated that they terminated drug testing because it was not cost beneficial.

For non-mandated companies, the majority, 92 percent, favor urine analysis over other forms of detection, such as blood, hair, and performance impairment systems. A discussion of urine analysis follows.

URINE ANALYSIS

The value and strengths of counter drug programs based on urine analysis are that the collection, control, analysis, and reporting procedures have been worked out and are reliable. Service industries exist to serve corporations, screening costs are low, and only urine testing positive needs further confirmational testing.

Urine analysis is, however, easily defeatable. Tests are avoided, employees have themselves pre-screened, they take diuretics and drink fluids in excess, switch samples, or adulterate their samples.

Pre-employment testing with urine is easily planned for or delayed by a user. Users know, or can find out, how many days they need to abstain from use of a drug in order pass the urine analysis test. Table 1. was compiled from the cited sources to show the period of time after drug use for likely detection in urine.

Table 1. Likely period after use for detection of illicit drugs in urine.

ILLICIT SUBSTANCE	DETECTION LEVEL	PERIOD OF LIKELY DETECTION
HEROIN	300 NG/ML	1-4 Days [9]
COCAINE	300 NG/ML	8-48 Hours [9]
MARIJUANA	100 NG/ML	7-34 Days [9]
LYSERGIC ACID DIETYLAMINE (LSD)	20 NG/ML	2 Days [5]
PHENCYCLIDINE (PCP)	75 NG/ML	5-10 Days [9]

Except for marijuana and PCP, most drugs metabolize rather quickly. This makes urine testing a very poor choice for screening for users in the work place unless the level of detection is lowered to a value near the lower detection limit of the Gas Chromatograph/Mass Spectrometer (GC/MS). The screening level used for cocaine is 300 ng/ml. This is far above the lower detection limit of today's laboratory instruments.

The immunoassay screening kits used to identify positive samples for GC/MS analysis are designed to detect concentrations at or above 300 ng/ml for cocaine. The combination of two independent, sensitive, and specific tests ensure the integrity of the urine analysis. Neither technique, immunoassay, or GC/MS, is recommended by itself as the basis for a conclusion on presence of an illicit substance or its metabolites.

The predictive value of positive urine analysis with GC/MS is a function of the population sampled [Uthman, 9]. Dr. Uthman pointed out in his paper that for populations with low prevalence of drug use, the predictive value for a positive indication for GC/MS is rather poor, and the predictive value becomes quite high for high percentage of drug abusers in the sample (See Table 2.).

Table 2. Predictive value of GC/MS as a function of prevalence of drug abuse.

Prevalence of Drug Abuse by Percent of Population	Predictive Value for a Positive
0.1 %	7.1 %
1.0 %	43.4 %
10.0 %	89.4 %
20.0 %	95.0 %
50.0 %	98.7 %

GC/MS is a poor predictor when the percentage of users is small relative to the total population. Dr. Uthman states that for a population of 1 user per 1000, 93 % of the positives reported will be false, and for populations of 200 users per 1000, only 5 % of the positives will be false.

Table 2 confirms that the DOT procedures for screening urine with immunoassay tests at higher mass concentrations than the GC/MS confirmational levels (Table 3.) are valid. First, the immunoassay is an independent technology with its own degrees of false positives and negatives. Second, the confirmational levels are run within the range of the immunoassay tests and the lower detection limit of the instrument. The threshold levels used by Dr. Uthman were unreported. Thresholds nearer the lower detection limit of the instrument would be expected to have higher false positives than thresholds that are usually designated for confirmational testing.

False negative rates with urine analysis are unknown. False negatives can be reduced by using the lower detection limit of the GC/MS instrument as the threshold rather than the DOT criteria (Table 3.). The lower detection limit of the instrument is the concentration level below which the instrument is unreliable. DOT UA thresholds are very lenient due to the need to maintain low false positives. Testing by the GC/MS at thresholds above the lower detection limit would most likely catch more users, but at the same time result in more false positives.

Table 3. Concentration levels required by DOT [10].

ILLICIT SUBSTANCE	INITIAL SCREENING	CONFIRMATIONAL GC/MS TEST
OPIATES (MORPHINE, CODEINE)	300 NG/ML	300 NG/ML
COCAINE	300 NG/ML	150 NG/ML
MARIJUANA	50 NG/ML	15 NG/ML
PCP	25 NG/ML	25 NG/ML
AMPHETAMINES	1000 NG/ML	500 NG/ML

DOT UA procedures are well established and accepted for counter drug programs in the work place. However, UA has shortcomings which need to be addressed. The shortcomings are that:

1. Users can be missed (for reasons stated above)
2. Productivity is lost due to sampling/collection
3. Sampling may be considered intrusive
4. Collection can be messy
5. Tests are too easily defeated to be considered a sufficient deterrent

OTHER SCREENING TECHNOLOGIES AND SAMPLE SOURCES

Screening technologies and sample targets, other than urine analysis, have been developed since 1988. These technologies include ion mobility spectrometry (IMS), immunoassay surface wipe kits, residue detecting sprays, and impairment testing. Sampling targets include surfaces, clothing, hair, sweat, and saliva.

Surface Sampling Technologies: Surface sampling systems include analytical instruments based on ion mobility spectrometry (IMS), immunoassay kits, and chemical sprays. Surface samplers are non-intrusive.

The IMS technology is very mature. IMS instruments capitalize on the fact that ions of different chemical substances transit in an electric field at characteristic and identifying speeds. Chemical substances leave residues that can be collected and readily analyzed with IMS instruments. These instruments can analyze for multiple compounds with every sample and complete an analysis within 5 seconds. IMS instruments can be trained to detect many substances.

IMS systems are used by the Drug Enforcement Administration , the Federal Bureau of Investigation, the U. S. Customs Service, and the U. S. Coast Guard. IMS instrumentation is also used for surface analysis in detecting explosives residues associated with terrorist bombs.

Portable IMS residue analyzers are now available commercially. Samples can be acquired from items that are touched or worn. Clothing is an excellent collector of drug particles and sampling of clothing is easily performed via vacuuming. Surface screening of items such as drivers' licenses, identification cards, purses, wallets, etc., can be performed fast and inexpensively. Positive results with surface sampling will identify which person has been in contact with illicit drugs. The individual could be a user, dealer, or have associated with people who either deal or use illicit drugs. Confirmation of drug use requires hair or urine analysis. For non-using dealers, investigative techniques, such as surveillance, would be called for.

Candidate targets for sampling and analysis include:

1. Clothing, time cards, lockers
2. Identification cards and drivers' licenses
3. Personal belongings (handbags, wallets, gym bags, radios, etc.)
4. Hands

Immunoassay kits are residue samplers that are based on the interaction between the substance of interest and an antibody conjugate. A chromatographic process identifies that conjugate binding has taken place. These immunoassay surface screening tools are sensitive to 5 - 10 nanograms and respond in about one minute. One substance can be tested with each immunoassay surface wipe. Presently, there are three relatively inexpensive, $15 per kit, drug residue kits: cocaine, marijuana, and heroin.

Colormetric surface sprays are sensitive to about one microgram for cocaine, heroin, and marijuana. The surface sprays use a swiping pad that picks up surface particles. The pad and the residue are sprayed, chemicals in the spray react with the substance and produce a unique color. For example, the chemical spray for cocaine consists of cobalt thiocyanate which turns turquoise when cocaine is present. Substances are tested for one at a time. Positive responses are obtained within seconds. The sprays are packaged sufficient to test 100 samples at $50 per can.

Hair testing. Hair testing has become an accepted method for screening individuals for illicit drug use, and is extremely useful for pre-employment screening. Hair collection is easy and relatively non-invasive. The samples can be stored, split for retest, or recollected for challenged results. Hair provides an historical (up to 90 days) use of drugs. Hair as a test matrix is resistant to tampering or adulteration [5].

The analysis of hair is performed by a number of commercial companies. The hair sample quantity varies by company and can be as large as the thickness of a pencil. The analysis consists of a pre-screening for cocaine, marijuana, heroin, LSD, and PCP. The screening is performed using immunoassay tests. Positives are then analyzed via GC/MS for confirmation. The results take about one week to acquire and cost about $48 per sample.

Impairment testing. Another technique being marketed for reasonable suspicion is performance impairment visual tests. Impairment detection systems can inexpensively detect impairment caused by alcohol, prescription drugs, illicit drugs, and sleep disorders. These systems require a previously compiled data file, presumably acquired when the person is known to be "clean," perhaps established during pre-employment screening. The compiled data based is then compared with subsequent "readings" on individuals.

The tests could be random, based on reasonable suspicion, or scheduled for activities that are highly safety sensitive (airline pilots, truck drivers, bus drivers, etc.). Employees who show impairment most likely will need to have urine screenings performed to determine the nature of the impairment. However, the results of impairment testing are sufficient to prevent impaired employees from performing safety sensitive jobs.

Sweat and saliva testing. Sweat and saliva are two additional matrixes for detecting illicit drug use. Both sweat and saliva are collected fairly non-intrusively. Sweat can be collected directly from the underarm, forehead, or through the use of sweat patches [8]. Sweat patches are placed on a person's arm and collect the sweat over a period of time; normally one or more days.

Sweat patches are similar to band aids. Some patches under development contain a chemical that induces sweating in the area under the patch, thus eliminating the need to have a patch worn for long periods. Patch samples can be split for testing. One split sample could be tested with a surface residue technology (IMS, immunoassay, or colormetric sprays) and the remaining split could be tested by a laboratory using GC/MS to confirm the results of the surface residue techniques. People whose sweat is positive from the surface testing could be directed for confirmation through urine analysis. UA is obligatory for mandated companies and recommended if any consequence is planned for an employee. Splitting samples is also valuable for purposes of reanalysis in the event that the results of the testing are challenged.

Drugs in sweat can come from two sources: the metabolic process and contamination. Reported results in measuring drug concentrations in perspiration were reported by Richards [7]. The results are shown in Table 4.

Saliva can be collected from a sterilized tongue depressor from the mouth piece of an alcohol breath analyzer test. The collected saliva again can be tested with immunoassay surface residue samplers. Positive results with surface test systems could be confirmed with UA. For mandated businesses, the saliva test results would be considered as a reasonable suspicion observation and UA is mandatory. Non-mandated companies planning some form of consequence would be advised to have the employee under go UA.

UA as a confirmation for either positive saliva or sweat tests is important, because UA is the accepted procedure by the courts. However, the UA needs to be performed at thresholds at the GC/MS lower limit of detection. This is because the immunoassay surface kits are more sensitive than the immunoassay kits used to screen urine. The policy statement needs to spell out these procedures.

Table 4. Measured results of drug concentrations in sweat.

DRUG	CONCENTRATION Micrograms/milliliter	RANGE Microgram/milliliter
Methamphetamine	1.4	0.88-1.42
Morphine	1.5	0.31-2.7
THC	0.32	0.034-1.0
Benzodiazepine	0.19	0.14-0.33
Cocaine	50	3.4-317
Barbiturate	70	66-74
Methadone	0.48	0.31-0.86
Cotinine (Nicotine metabolite)	0.51	0.10-0.93

SCENARIOS FOR USE OF VARIOUS TECHNOLOGIES

Work place drug screening consists of the following scenarios:

1. Pre-employment Screening
2. Reasonable Suspicion Screening
3. Post Accident Testing
4. Random Testing
5. Return to Duty Testing
6. Follow-up Testing

Pre-employment Screening. The objective of pre-employment screening is to ensure that new employees are not drug users or abusers. The applicability of the various technologies for pre-employment screening is discussed below.

Surface residue detection systems: IMS, immunoassay kits, or colormetric sprays have limited use in pre-employment screening. These systems could be ultilized to screen application forms, etc., but positive indications would need to be confirmed by another technology such as urine or hair analysis. Drivers' licenses and other identification documents would be more ideal targets.

The question of invasion of privacy might limit the surface tests to corporate documents, which by themselves are a poor sampling matrix since these documents are generally handled by prospective employees for short time periods.

Urine analysis is inadequate for meeting the pre-employment objective. The prospective employees most likely will be informed that urine analysis will be performed. This gives the prospective employee time to abstain or to plan to subvert the test by adulteration or substitution.

Impairment testing requires a baseline performance file for each person. The generation of this file requires that the person, whose baseline is being gathered, is known not to be impaired. This requires on-the-spot urine analysis. Thus the urine analysis results would be the determining test for pre-employment, not the impairment testing. The collection of the baseline data for later impairment testing is, however, recommended at this time.

Hair analysis is the most appropriate drug screening tool for pre-employment testing. The hair retains the metabolites of illicit drugs for up to ninety days. Forewarning the prospective employees does not give them the opportunity for counter measures as with urine analysis. Positive results are sufficient for terminating the employment process.

Reasonable Suspicion Screening. The purpose of Reasonable Suspicion Screening is to allow the supervisor to respond to employees whose behavior, odor, or appearance is indicative of potential drug use. Trained supervisors can request these employees to undergo a drug test.

Surface residue analyzing systems (IMS, immunoassay kits, and colormetric sprays) are useful in extending a supervisor's subjective observation capabilities. These technologies provide the opportunity to use objective information for reasonable suspicion. Employees whose tools, time cards, phones, lockers, corporate vehicles, etc., are found to contain illicit drug residues could be requested to test for illicit drugs. High safety risk work areas, devices, tools, and vehicles could be sampled to pin point reasonable suspicion. Sampling with IMS instruments is fast, inexpensive, and can detect several drugs per sample. Immunoassay residue samplers are limited to one illicit drug per sample analysis. These are more expensive and are probably best used as confirmation to support IMS findings or intelligence information about an employee and a specific drug.

Impairment testing of reasonable suspicion candidates is a valid method of determining "fit for duty" status. Employees identified by the trained supervisor could be tested with an impairment performance system. Test results that are positive would be grounds for removing the employee from a safety sensitive job. Confirmation of this for DOT regulated industries would still require urine analysis. For non-DOT companies, further testing for confirmation would be unnecessary as long as this is spelled out in the company's drug policy statement.

Hair analysis is limited because of the delay time for a laboratory analysis (about one week).
Urine analysis can be performed on-site or the employee can be sent to a specimen collection site or laboratory for more immediate results.

Post Accident Testing. For DOT mandated companies, covered employees involved in accidents must be urine tested. For non-mandated companies, surface sampling could provide useful information on the possibility of illicit drugs involvement. Hair analysis would confirm or deny that the individuals involved in the accident had or had not been using drugs in the past, but is unable to confirm use at the time of the accident. Impairment testing immediately after the accident could determine whether the employee was on illicit or licit drugs, alcohol, or was experiencing some kind of sleep disorder. Urine analysis is the most confirmational post accident test.

Random Testing. DOT mandated companies must use urine and test 50% of employees each year. Mandated companies could use other technologies on a random basis, but only in the context of reasonable suspicion observations, not in the context of random testing. Non-mandated companies can use other technologies randomly. Surface samplers can be employed and any positive result confirmed with urine or hair analysis. IMS surface residue samplers could be useful for periodic, random, sampling of the work place for illicit drugs. Impairment testing could also be used and positive results confirmed with urine.

Return to Duty Testing. DOT mandated companies are to use urine testing to determine if an employee can be returned to duty. Non-mandated companies can use urine, hair, surface samples, or impairment tests.

Urine testing is an insufficient test due to the problems mentioned earlier. Employees will very likely know that they are to be tested on a specific date. Hair analysis is the most certain test to verify that the employee has been drug free for a long period of time prior to resuming duties. Impairment testing could be used very effectively for fast, inexpensive testing in return to duty situations for non-mandated companies. Surface analyzers have limited utility for this screening need. The presence or absence of residue is an insufficient indicator that drugs are being used by employees. Confirmation via urine, hair, or impairment testing is needed.

Follow-Up Testing. DOT requires that random urine tests be spread out over a course of one calendar year for employees returned to duty after failing a previous drug test. For non-mandated companies random sampling of the surfaces in the employees' work area is a very effective, inexpensive means of monitoring for potential drug abuse. The IMS systems allow for sampling and testing for many drugs with each surface sample. In the event that an employee switches to another illicit drug from the one originally found, this switch can be detected. The use of immunoassay systems are valid for employees who may not switch to other substances. Hair and impairment testing are very strong systems for use in follow-up testing.

THE KEYS TO A DRUG-FREE WORK PLACE

A successful counter drug program requires the following seven key activities. The degree to which they are performed may vary by corporate size, but each key will help build a successful program. The seven keys are:

I. Define corporate objectives:
    A. Achieve mandated requirements, if covered by DOT regulations
    B. Define measures of effectiveness for program
        1. Liability insurance changes
        2. Productivity changes
        3. Theft and embezzlement losses

4. Accident rates
5. Workman's compensation expenses
6. Absenteeism and tardiness rates
7. Quality of Inter-employee relations

II. Assess policy in effect
  A. Estimate present level of drug abuse
  B. Inform employees on findings
  C. Get inputs from employees and supervisors
    1. Surveys
    2. Interviews
    3. Focus Groups

III. Develop plan to eliminate problem
  A. Weigh various drug screening options: risks versus benefits
  B. Who, when, frequency, and type of testing appropriate for scenario
    1. Pre-employment, reasonable suspicion, random, accidents, return to duty
    2. Hair, Urine, Surface Residue, Impairment

IV. Write and disseminate policy and plan to all potential and current employees
  A. Define consequences
  B. Define process for employees to contest results

V. Implement zero tolerance plan

VI. Track measures of policy effectiveness (Compare with data acquired in I. B. 1-7.)

VII. Periodically determine level of drug abuse among employees
  A. Review statistics on positive versus negative findings
  B. Assess benefit of program
  C. Determine if changes to program are needed

BIBLIOGRAPHY

[1] 1996 AMA Survey Workplace Drug Testing and Drug Abuse Policies, Summary of Key Findings, Research Report, American Management Association, 135 West 50th Street, NY, NY 10020-1201

[2] Kidwell, D. A., Ph.D., "The Alternative Matrix Program for Drug Abuse Detection and Deterrence", ONDCP/CTAC Drug Abuse Treatment Technology Workshop August 1995, Proceedings, p.3.13-3.36

[3] Mumm, Rosemary, MS, "The Orleans Parish District Attorney's Diversionary Program", ONDCP/CTAC Drug Abuse Treatment Technology Workshop August 1995, Proceedings, p.3.1-3.12Uthman, E. O. Diplomate, American Board of Pathology, April 1993

[4] National Institute of Justice, "A Criminal Justice System Strategy for Treating Cocaine-Heroin Abusing Offenders in Custody," U.S. G.P.O. 1988-202 Olt518OO82.

[5] Papac, D. I. & Foltz R. L., "Measurement of lysergic acid dietylamine (LSD) in human plasmas by gas chromatography/negative ion chemical ionization mass spectrometer", Journal of Analytical Toxicology, V14n3, May-June 1990, p. 189-190.

[6] Pilon, P. Dr., Hupe', "Document scanning as an Effective Method for Narcotics Interdiction", Proceedings, Counterdrug Law Enforcement: Applied Technology for Improved Operational Effectiveness International Technology Symposium, October 24-27, 1995, Part 1, p. 9.7-9.25

[7] Richards, G. F., JPL/CalTech, "Telemetered Drug Detection System: A Demand Reduction Tool", ONDCP/CTAC Drug Abuse Treatment Technology Workshop August 1995, Proceedings, p.3.37-3.52

[8] Swan, N., NIDA Notes Contributing Writer, "Sweat Testing May Prove Useful in Drug-Use Surveillance", NIDA notes, September/October, 1995, [NIDA Home Page][NIDA Notes Index][NIDA Notes 1995 Achive]

[9] Uthman, E. O., Diplomate, American Board of Pathology, April 1993

[10] Code of Federal Regulations, 1994, 40.29

Detection of drug usagc via breath analysis with an immunoassay film badge

H. Richard Lukens

Diametrix Detectors, Inc.
8221 Arjons Dr., Suite F, San Diego, CA 92126

ABSTRACT

A monolayer of antibody on a semimirror comprised of small islands of indium acts as a sensor capable of detecting vapors at extremely low concentrations without the use of wet chemistry. Already shown capable of detecting cocaine vapors at 4 femtograms per cc of air, the use of the device, called an immunoassay film badge, for detecting drugs on the breath is a natural extension of the sensor's use. This paper describes this application and initial experiments that demonstrate its feasibility.

Keywords: immunoassay film badge, dry immunoassay, vapor detection, breath analysis

1. INTRODUCTION

Breath analysis, which is typically carried out in a research environment, has potential usefulness in toxicology and in the diagnosis of disease. Examples are Manolis[1], who discusses breath analysis in diagnoses, and Stewart[2], who examines some toxicologic applications. The key point of breath analysis is that the concentration of a substance in the breath is related to its concentration in blood. One application of breath analysis that exists outside of the research environment is the determination of blood alcohol. Alcohol's partition ratio (the ratio of breath volume to blood volume to contain a given amount of alcohol) has been determined to be about 2100 (Dubowski[3]). Hence, the determination of alcohol concentration in the breath is a quantitative determination of blood alcohol concentration.

Except for the determination of blood alcohol, the equipment used for breath analysis depends heavily on gas chromatography (GC), mass spectrometry (MS), or a combination thereof. Breath traps (chemical, cryogenic, or adsorptive), which are necessarily to concentrate analytes and, also, to remove interfering subtances (particularly water), are used on the front end of the GC, MS, MS/MS, and GC/MS equipment, as described by Phillips [4]. O'Neill et al report that such instruments have detected over 380 compounds in human breath, including some compounds of fairly low vapor pressure[5]. For example, Manolis[1] indicates that tetrahydrocannibinol (THC) is detectable on the breath. THC has a vapor pressure of 5E-9 atm. at 22 degrees C [6].

Phillips[4] points out that there is a pressing need for simplification of breath analysis equipment, particularly with respect to implementing its practical applications. Since, the immunoassay film badge of Diametrix Detectors Inc. (DDI) is a compact, sensitive, and specific device that does not require a preconcentration or other sample preparation step, it was decided to investigate its potential application as a breath analysis tool.

2. The Immunoassay Film Badge (IFB)

The use of an indium semimirror, coated with a monolayer of protein for the detection of antibody toward the protein, in solution, was demonstrated by Giaever[7]. The converse, use of a monolayer of antibody on the indium semimirror to detect the antibody's target substance in solution , was demonstrated by Lukens and Williams[8] for both a protein (bovine serum albumin) and a small organic pesticide residue (2-aminobenzimidazole). Then Lukens and Williams[9] showed that the sensor comprised of antibody on an indium semimirror could also detect vapors of a target substance in air.

Operation of the IFB depends on a combination of physics and biochemistry. The semimirror does not conduct electricity, because the indium exists as non-contiguous islands on its clear substrate (glass or plastic). A basis for understanding the operation of the IFB as a surface plasmon phenomenon is provided by Sambles[10]. The angular distribution of light transmitted through, and reflected from, the semimiror depends on the light-stimulated oscillations of the islands' surface free-electrons, and that interaction is modulated by the presence of a dielectric coating. For this reason, the optical density (OD) of the sensor is greater than that of the bare semimirror, when measured by direct (normal) light. The modulation is a function of dielectric thickness, so when the sensor binds its target substance, the OD is measurably increased further.

As can be seen, the basic IFB system consists of a semimirror, to which is attached a monolayer of antibody toward the target substance, and a photometer.

When the sensor is exposed to air, it is possible that some environmental influence (e.g., dust) might also increase the OD. The IFB is formulated to compensate for such an eventuality. The IFB has 3mm diameter monolayers of immunoglobulin G (IgG) on the semimirror, one of which serves as a comparator, since it is made with an IgG that will not bind anything likely to exist as a vapor in the environment. Each 3mm area of IgG is called a spot, since it is easily visible to the eye. The other spots are made with antibodies toward target vapors. A 22 x 22 mm mirror can easily accomodate 4 spots, while a 1 x 3 inch mirror can hold up to 20 spots. That is, the latter size can be used to make IFBs toward a panel of 19 different substances.

An immunoassay device, the IFB has the exquisite specificity afforded by antibodies. It has been shown to successfully detect the room temperature vapors of numerous substances, including some with vapor pressures in the low trillionths of an atmosphere (cocaine, morphine, PETN, and RDX)[11]. It has also been shown to operate successfully over the temperature range of 1 to 51 degrees C and in relative humidities of from 10 to 100 percent in the detection of cocaine[12]. The sensitivity of the device is exemplified by the fact that the saturation concentration of cocaine in air at 1 degr. C is about 4 femtograms/cc. The ability to operate above ambient temperatures and at high humidity favors the breath analysis application.

The high levels of sensitivity and specificity of the IFB eliminate the need for sample preconcentration and processing. The antibody selects and binds target vapors directly from the air, with a concomitant increase in sensor OD. Thus, the ratio, R, of ODs, sensor/comparator, increases when the sensor binds its target molecules.

The ability of the IFB to operate above ambient temperatures and at high humidity seemed to favor the breath analysis application. It was decided to use caffeine as the target substance in testing the feasibility of using the IFB to detect a substance on the breath.

At the present time Diametrix Detectors, Inc. (DDI) is conducting a program, supported by the U.S.Customs Service, to improve the uniformity of IFB performance. Sources of non-uniform performance have been found to be variabilities in semimirrors and antibodies. Although, the former variability has been largely eliminated, quality control measures are applied to each batch of mirrors. An in-house quality control procedure to supplement those of antibody vendors is being put in place after it was discovered that vendor procedures are somewhat fallible.

3. EXPERIMENTAL

3.1 Equipment and Procedure

Monoclonal antibody toward caffeine was obtained from Biodesign International. Its affinity constant for caffeine was given as 2.1E9, and it was specified as having a slight cross reactivity toward several other methylated xanthines. Non-specific bovine IgG, lyophilized, was obtained from Sigma Chemical Co. to serve as the comparator. Each antibody was prepared as a dilute (0.2mg/ml) solution in slightly alkaline saline solution with 0.02%w azide preservative.

Indium semimirrors were prepared by evaporation of indium, under high vacuum, from a resistively heated tantalum boat onto cleaned 22mm x 22mm microscope slide cover glasses by Surface Optics Corporation to DDI specifications.

A Macbeth model 931 densitometer, equipped with a tungsten light source and capable of reading optical density (OD) to 3 decimal places, was used to measure the ODs of IFBs. The instrument was modified by installing a graduated microscope stage for holding and accurate placement of specimens. The instrument was zeroed and calibrated, using authenticated gray scales, prior to each measurement session.

The procedure for making an IFB was as follows. Aliquots, 0.05 ml, of diluted antibody solution were placed at defined spots on a semimirror and allowed to incubate for 10 minutes at room temperature. The aliquots were mostly removed with transfer pipets, and each spot dynamically rinsed with distilled water with care to avoid mixing of antibodies or rinse water from adjacent spots. The IFB was then allowed to dry at room temperature overnight.

Four spot IFBs contained 3 sensor spots (antibody toward a specific substance) and one comparator spot. Two spot IFBs had one each sensor and comparator spots.

All IFBs were prepared in a laboratory having positive air pressure and an outside air supply, features that make it unlikely that the room will contain vapors of the target substance.

The procedure for testing an IFB's ability to detect its target vapor was as follows: pre-exposure OD measurements of each spot were obtained. The IFB was then exposed to test vapors, after which the OD of each spot was remeasured.

Three methods were used to expose IFBs to test vapors, two of which involved 100 ml Coplin staining jars equipped with screw-on caps. One jar had an open 1 dram vial containing 4 grams of coffee. The jar was kept closed for a period of at least 24 hours before inserting an IFB, in order to assure saturation of caffeine vapors. Placement of an IFB in the jar was carried out with care to minimize turbulence. The jar was then closed to allow exposure to caffeine vapor for a measured period of time.

A second jar had only an IFB, and was used for breath testing as follows: ten cups of coffee were brewed with 65 grams of ground coffee. The author consumed a cup of coffee over a period of 30 minutes, and after an additional 5 minutes, breathed a measured number of times into the jar. Only the latter half of each breath was directed into the jar, since it is the portion of breath most in equilibrium with blood circulating through the lungs. The jar was then closed for a measured time, after which the ODs were again measured.

The third method was to consume coffee, as described above, then direct the latter half of a measured number of breaths across an IFB. In this last method, the IFB was not allowed to dwell in the presence of breath beyond the time it took to breath on it.

In addition to the IFBs for detecting caffeine, a semimirror with two comparator spots was prepared, and the OD ratio between the spots was measured at sequential times over several days, in order to establish the background fluctuation in OD ratio measurements.

3.2 Results

The various IFBs that were exposed to caffeine vapors from ground coffee and/or breath are described in Table 1, together with the length of exposure and the change in OD ratio (senor OD/comparator OD), in Table 1.

Table 1. Exposures & Results

IFB	Sensor Spots	Exposure Number	Exposure Type	Incubation Time, min.	Change in OD Ratio (a)
1	x	1	coffee	1000	0.98
2	x	1	breath	1000	0.47
3	x, y, z	1	breath	180	0.57, 0.83, 0.47
4	x, y, z	1	breath	60	0.53, 0.28,-0.13
5	x, y, z	1	breath (b)	0	0.39, 0.22,-0.03
		2	breath (b)	0	0.41, 0.39,-0.03
		3	breath (b)	0	0.42, 0.41,-0.04
		4	coffee	1100	0.57, 0.60, 0.31

(a) For IFBs with three sensor spots, the values are given for spots x, y, and z, respectively. The values given are the percent, relative, change in the ratio. The intial ratio, in all cases, was close to unity.
(b) In each of exposures nos. 1 and 2, the last halves of three breaths were directed across the IFB. In exposure no. 3, the latter half of six breaths were directed across the IFB. The time to direct each breath across the IFB was about 5 seconds.

The instrumental background variation in the OD ratio was established to be 0.00 with a standard deviation of +/-0.16. Thus, the 16 positive values in the last column of Table 1 represent the detection of caffeine beyond the 95% level of confidence, and the four negative values fall within the range of the null response.

4. DISCUSSION

The anti-caffeine antibody was found to be satisfactory, as shown in the first IFB of Table 1, which gave a strong response upon exposure to ground coffee. Coffee has on the order of 1%w caffeine, and the procedure used assured the presence of caffeine vapors in the Coplin jar.

The strong plurality of positive tests for caffeine with IFBs exposed to breath, after the consumption of coffee, indicates that such a simple approach to breath testing is feasible. Given equal quality of semimirrors and antibody, the IFB will be equally effective in the determination other compounds on the breath.

The results from IFBs numbers 4 and 5 in Table 1 show that, in the IFB's present state of development, reduction of the chance of a false negative result can be achieved through the use of more than one sensor spot toward the target compound. As previously indicated, DDI is presently working toward improvement in IFB uniformity, which will further reduce the chance of a false negative. The chance of a false positive result is minimzed by the specificity of the antibody and the use of a comparator.

Because each sensor spot of an IFB has a finite number of target-compound binding sites per unit area, it is potentially a quantitative device. When uniformity of IFBs is achieved, it will be possible to take advantage of this property. In particular, the fraction of the maximum change in OD ratio, which is achieved when all the sites are occupied, is a function of the time of exposure and concentration of target vapor. This is seen to some extent in the case of IFB number 5, in Table 1, where 53%, 68%, and 71% of the saturated OD change for that IFB is obtained with a total of 3, 6, and 12 breaths, respectively. Given a uniform and calibrated IFB behavior, the OD change at a given time will indicate the concentration of target vapor.

The data in Table 1 also indicate that the time to minimmally detect caffeine was less than 5 seconds.

Additional requirements for determining the blood concentration of a target compound via breath analysis, beyond a uniformly behaved and calibrated IFB, are knowledge of the vapor pressure of a target compound and its partition ratio.

5. ACKNOWLEDGEMENTS

This research is part of a program supported by the U.S.Customs Service. The author wishes to thank Amanda Wu, of DDI, and Sam Dummer, of Surface Optics Corporation, for their contributions to the program. Ms. Wu, who voluntarily eschewed beverages and foods with a caffeine content, prepared the caffeine-sensing IFBs and measured their ODs. Mr. Dummer prepared the semimirrors.

6. REFERENCES

1 A.Manolis, "The Diagnostic Potential of Breath Analysis," Clin.Chem., vol.29, pp.5-15, 1983.

2.R.D.Stewart, "The Use of Breath Analysis in Clinical Toxicology," in Essays in Toxicology, vol.5, W.J.Hayes, Jr., ed., Academic Press, 1974.

3. K.M.Dubowski, "The Technology of Breath-Alcohol Analysis," U.S.Dept. of Health and Human Serivices Publication No. (ADM)92-1728, 1991.

4. M.Phillips, "Breath Tests in Medicine," Scientific American, pp.74-79, July, 1992.

5. M.J.O'Neill, S.M.Gordon, M.H.O'Neill, R.D.Gibbons, and J.P.Seidon, "A Computerized Classification Technique for Screening for the Presence of Breath Markers in Lung Cancer," Clin.Chem., vol.34, pp.1613-1618, 1988.

6. H.R.Lukens, unpublished data.

7. I.Giaever, "The Antibody-Antigen Reaction--A Visual Observation," J.Immunology, vol.110, pp. 1424-1426, 1973.

8. H.R.Lukens and C.B.Williams, "A Solid Substrate Immunological Assay for Monitoring Organic Environmental Contaminants," EPA Report EPA-600/1-77-018, 1977.

9. H.R.Lukens and C.B.Williams, "Method for Detection Organic Vapors," U.S.Patent No.4,353,886, 1982.

10. J.R.Samble, G.W.Bradbery and F.Yang, "Optical Excitation of Surface Plasmons: and Introduction," Contemporary Physics, vol.32, pp.l173-183, 1991.

11. H.R.Lukens, "An Immunoassay Film Badge for the Detection of CBW Agents and Explosives," in Proc. of the Ninth Annual Joint Government-Industry Symposium and Exhibition on Security Technology. American Defense Preparedness Assn., Security Technol.Div., pp.186-189, 1993.

12. H.R.Lukens, "Detection of Narcotics with an Immunoassay Film Badge," Proc. 34th Annual Meeting Nuclear Materials Management, Institute of Nucl.Matls.Mgmnt., pp.594-598, 1993.

Breath alcohol, multi sensor arrays and electronic noses.

Nils Paulsson[1] and Fredrik Winquist[2]

1. S-SENCE and SKL - National Laboratory of Forensic Science, S-581 94 Linköping, Sweden.
Email : nilpa@ifm.liu.se
2. S-SENCE and Applied Physics, Linköping University, S-581 83 Linköping, Sweden.
Email : frw@ifm.liu.se

ABSTRACT

The concept behind a Volatile Compound Mapper (VCM), or electronic nose, is to use the combination of multiple gas sensors and pattern recognition techniques to detect and quantify substances in gas mixtures. There are several different kinds of sensors which have been developed during recent years of which the base techniques are conducting polymers, piezo electrical crystals and solid state devices. In this work we have used a combination of gas sensitive field effect devices and semiconducting metal oxides. The most useful pattern recognition routine was found to be Artificial Neural Networks (ANN), which is a mathematical approximation of the human neural network.

The aim of this work is to evaluate the possibility of using electronic noses in field instruments to detect drugs, arson residues, explosives etc. As a test application we have chosen breath alcohol measurements. There are several reasons for this. Breath samples are a quite complex mixture containing between 200 and 300 substances at trace levels. The alcohol level is low but still possible to handle. There are needs for replacing large and heavy mobile instruments with smaller devices. Current instrumentation is rather sensitive to interfering substances.

The work so far has dealt with sampling, how to introduce ethanol and other substances in the breath, correlation measurements between the electronic nose and headspace GC, and how to evaluate the sensor signals.

Keywords: Electronic Noses, Alcohol, Breath, Gas Sensors

SPIE Vol. 2932 • 0-8194-2334-3/97/$10.00

2. ELECTRONIC NOSES

An electronic nose[1,2] or Volatile Compound Mapper (VCM) is a device which, in a very general sense, is built of similar building blocks as the human nose. A chemical sensor array responds to a gaseous substance, the response is evaluated by a pattern recognition routine and the result is expressed in known terms of qualification and quantification.

2.2 Multi sensor arrays

A multi sensor arrays is an array of sensors where each sensor has an often low but different selectivity from the other sensors. Each sensor adds a piece of new information to the system making the whole array a sensor system with high selectivity. An array of 10 binary sensors (either high or low output) with different selectivity would in the perfect case be able to classify 2^{10} (1024) gaseous substances in a mixture.

2.3 Sensor techniques

During years several different gas sensor technologies have been developed[3]. In this work two major sensor types have been used, MOSFETs with catalytic gates[3] and semiconducting sensors with metal oxide dopands (Tagushi)[4].

A MOSFET sensor is basically a field effect transistor where the transistor gate has been coated with a thin, porous layer of a catalytic metal. When the gate is exposed to an atmosphere containing combustible gases the catalytic metal adsorbs and oxidizes these gases. Charged reaction intermediates will be formed which induce dipoles, changing the electric field at the transistor gate and thus the electric behavior of the transistor (fig. 1).

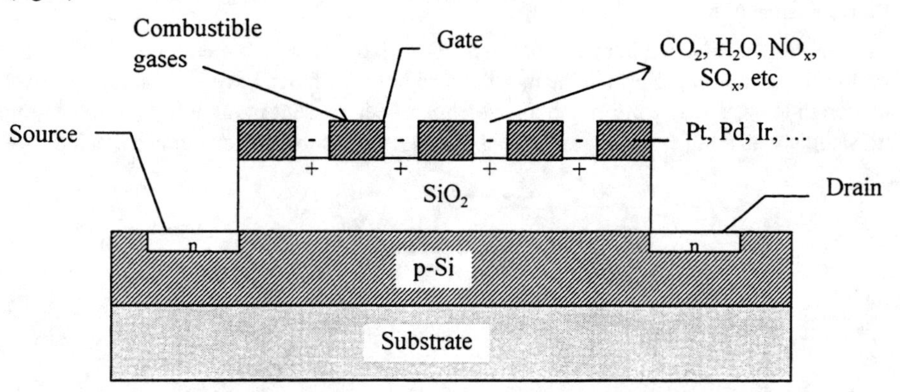

Fig 1. Gas sensitive MOSFET device with catalytic gate.

By using different metals with different surface structures, sensors with different selectivities can be manufactured. Commonly used metals are Pt, Pd, Ir and alloys thereof. Different surface structures are obtained by altering the metal film thickness and the coating conditions. The sensors are always operated at elevated temperatures (>100 °C) to prevent condensation of water. Since the selectivity also is influenced by the sensor temperature the same sensor can be operated at different temperatures with further increased discriminating capabilities of the sensor array as a result.

A Tagushi sensor is a solid state semiconductor mainly composed of sintered tin dioxide which detects gases through an increase in electrical conductivity. When exposed to an atmosphere containing reducing gases the tin dioxide adsorbs these gases and cause oxidation. This lowers the electron potential barrier between the sintered grains allowing electrons to flow more easily, thereby reducing the electric resistance.

2.4 Data evaluation

The most often used sensor evaluation algorithm is feed forward, back propagation artificial neural network (ANN[5]). Other algorithms used is ARX (system identification)[6], principal component analysis and partial least square. Which algorithm to use is heavily dependent on the actual VCM-application. The reason for the frequent use of ANN is the nonlinear behavior of sensor responses.

The common approach of these methods is that test measurements with known stimuli and responses are used to build linear or nonlinear models of the analysis setup. By using sensor data from many samples with known properties, e.g. ethanol concentration in breath samples, the algorithms try to describe the sensors behavior, i.e. build a model. This data set from samples with known properties is usually called learning set. The found model can be used to predict properties of new samples which have not been exposed to the sensor array before. The quality of the model depends on how extensive the learning set is. The larger learning set the better prediction capabilities of the model.

3. BREATH ALCOHOL ANALYSIS

The goal of this project is to evaluate the possibilities of using VCM-techniques in forensic field methodology. Breath alcohol measurement has been selected as test application. There are several reasons for this. Breath samples are quite complex mixtures containing between 200 and 300 substances at trace levels. The alcohol levels are low but still belong to one of the main components when present. It is relatively simple to collect many realistic samples. As a potential field method it is a also strong competitor to existing instrumentation concerning mobility and probably also selectivity.

3.2 Sample collection

Breath samples are collected in plastic bags (fig 2) with one inlet and one outlet. The test person holds the breath for 30 seconds and blows in the mouth piece until the lungs have been emptied. By letting the breath flow through the plastic bag the breath air which has been present in the lungs the most time is collected. A breath sample with a high concentration of emitted gaseous substances is thus collected.

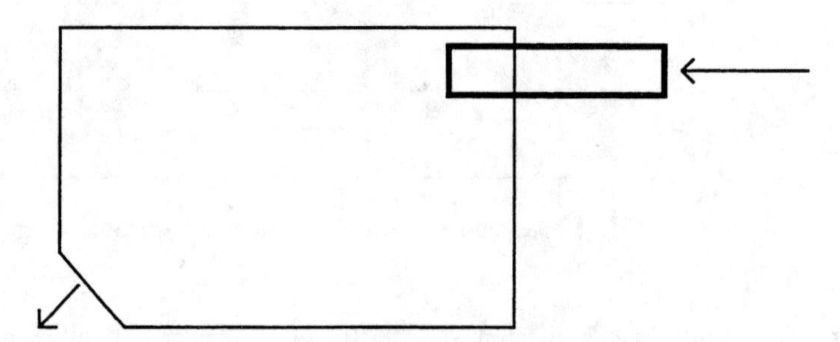

Figure 2. Plastic bag for collecting breath samples.

3.3 Alcohol in breath

Two different approaches have been used to introduce ethanol in breath samples. One way is to drink ethanol. This is clinically the preferred method but owing to some practical limitations, i.e. constantly drunken coworkers, sniffing is used as an alternative method. Just before the test person is about to blow in the plastic bag, he or she inhales the headspace from a beaker with 95% Ethanol. The concentration in the sample very well matches typical levels obtained from drinking alcohol.

3.4 Headspace GC

To be able to build a learning set it is necessary to know the actual ethanol concentration in the samples. This is obtained by headspace gas chromatography (GC). 400 ul headspace from the sample is injected in the GC and the ethanol peak area is calibrated towards reference gases of 75, 125, 250 and 500 ppm ethanol in argon.

4. BREATH ALCOHOL MEASUREMENTS

A test was performed to analyze ethanol metabolism by breath samples. A female test person consumed (< 1 min) 25 ml 95% ethanol (spiritus fortis) diluted in 2 dl black currant juice. A breath sample was taken every 10 minutes starting 10 minutes after the intake. A reference sample was also taken 10 minutes before. The samples was exposed to the VCM and the actual ethanol concentration was measured by gas chromatography. The sensor array consisted of 10 MOSFET sensors (table 1) and one infrared (IR) CO_2-detector. No Tagushi sensors was used in this test owing to technical problems. Gas samples were pumped from the plastic bags by a membrane pump at a calibrated flow rate of 50 ml air/min and injected in the sensor chamber during 45 seconds. The injection time was controlled by a PC operated gas valve.

Sensor Number	Sensor Class	Device Type	Operating Temperature (°C)	Detected Compounds
1	MOSFET	Pd, 6 nm	140	Hydrogen, amines
2	MOSFET	Pd, 35 nm	140	Hydrogen
3	MOSFET	Pt, 9 nm	140	Amine, aldehyde, alcohol
4	MOSFET	Ir, 9 nm	140	Amine, ester
5	MOSFET	Pt/Pd, 10 nm	140	Hydrogen, amine
6	MOSFET	Pd, 6 nm	170	Hydrogen, amine, aldehyde, ester, alcohol, ketone
7	MOSFET	Pd, 35 nm	170	Hydrogen
9	MOSFET	Ir, 9 nm	170	Amine, aldehyde, ester, alcohol, ketone
10	MOSFET	Pt/Pd, 10 nm	170	Hydrogen, amine, alcohol
12	IR			Carbondioxide

Table 1. Sensor array configuration.

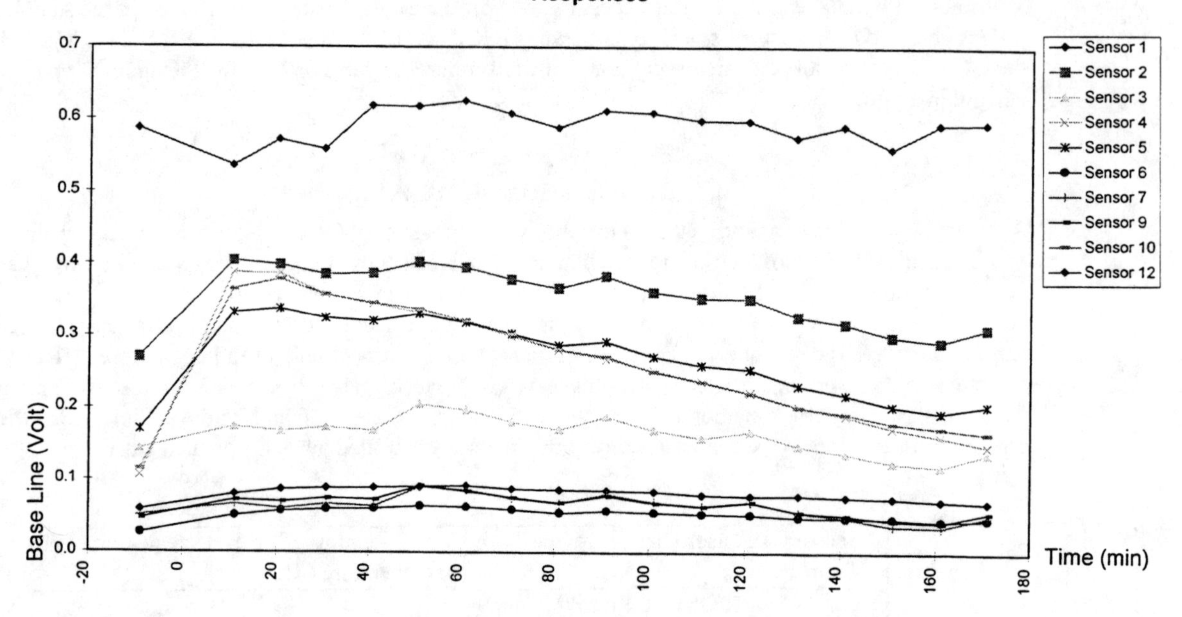

Figure 3. Sensor responses from breath alcohol samples.

Totally 18 samples were taken during a time period of almost 3 hours (fig 3). The sensor signals from the samples were merged with the corresponding ethanol concentration to form a learning data set (10 samples). The remaining 8 samples where used as validation data set. An artificial neural network (ANN) was designed to match the sensor signals with the ethanol concentrations obtained by GC. Figure 4 is a time plot of the actual concentration and the by ANN estimated concentration. The plot includes both the learning set and the validation data set.

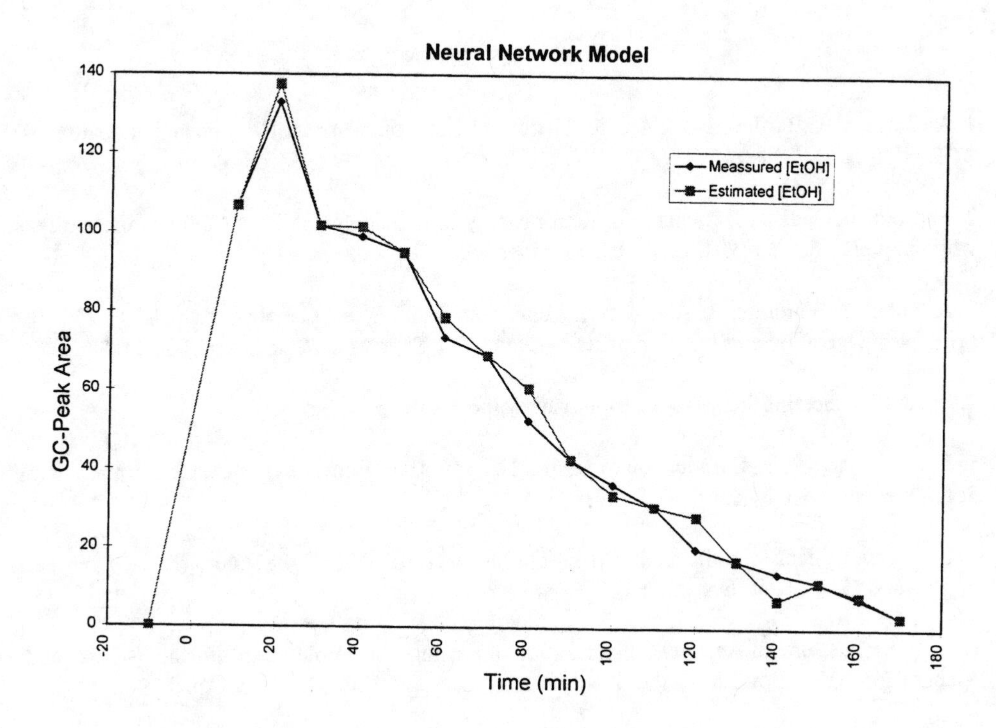

Figure 3. Actual and estimated Ethanol concentration in breath samples over time.

As can be seen in figure 3 it takes about 20 minutes before the highest ethanol concentration is reached. The delay depend on how long time it takes for the ethanol to traverse from the stomach to the blood[7]. What is noteworthy is that even though the learning set only contains 10 samples, the estimation capabilities of the ANN is still very good. The usual size of a learning set starts at some 50 samples and extends to a few hundred.

6. CONCLUSION

As the test above shows, the current VCM setup seems to be a promising analysis instrument concerning breath alcohol measurements. Not mentioned before is that the MOSFET sensors are rather slow in recovery. It takes less than a minute to get an informative signal but the sensors needs about 6 minutes to reach accurate base line stability. This time might however be decreased by using effective signal processing and enhanced sensor construction.

The next step in this work is to collect larger learning sets covering variations in breath samples from the same person over time and variations between different persons. The origin of these variations are probably due to food intake, environment and health status. Fasting for example results in an increased amount of exhaled acetone[7]. These tests might also indicate that it is possible to perform VCM based medical diagnostisation. There will also be a test performed to verify the behavior of neural networks when exposed to positive breath samples originating from ethanol consumption and ethanol sniffing.

REFERENCES

1. J. Gardner, P., Bartlett, 1994, A Brief History of Electronic Noses., Sensors and Actuators, B18-19, 211-219.

2. Sensors and Sensory Systems for Electronic Nose, Proceedings of the NATO advances research workshop, Reykjavik, 1991, Eds. J.W., Gardner, P.N., Bartlett.

3. A. Spetz, F. Winquist, H. Sundgren, I. Lundström, Field Effect Gas Sensors, p. 219 - 279, Gas Sensors. Principles, Operation and Developments / edited by G. Sberveglieri., ISBN 0-7923-2004-2.

4. Figaro Engineering Inc., 1-5-3 Senbanishi, Mino, Osaka 562, Japan.

5. J., Lawrence, 1992., Introduction to Neural Networks and Expert Systems., California Scientific SoftWare, Nevada City.

6. L., Ljung, System Identification. Theory for the User., ISBN 0-13-881640-9, P T R Prentice-Hall Inc., Englewood Cliffs, New Jersey 07632.

7. A. W., Jones, Breath Acetone Concentration in Fasting Male Volunteers: Further Studies and Effect of Alcohol Administration., Journal of Analytical Toxicology, Vol. 12, March/April 1988.

Design and Field Results of a Walk-Through EDS

Gregory J. Wendel, Edward E. A. Bromberg, and Memorie K. Durfee
Thermedics Detection, Inc., 220 Mill Road, Chelmsford, MA 01824

William Curby
USFAA Technical Center, Atlantic City International Airport, Atlantic City, NJ

Abstract

A walk-through portal sampling module which incorporates active sampling has been developed. The module uses opposing wands which actively brush the subjects exterior clothing to disturb explosive traces. These traces are entrained in an air stream and transported to a High Speed GC-chemiluminescence explosives detection system. This combination provides automatic screening of passengers at rates of 10 per minute. The system exhibits sensitivity and selectivity which equals or betters that available from commercially available manual equipment. The system has been developed for deployment at border crossings, airports and other security screening points. Detailed results of laboratory tests and airport field trials are reviewed.

Keywords: explosives detection, chemical sensing, gas chromatography, portal, human inspection, trace detection.

Introduction

Currently there are several trace explosive detection systems on the market which rely on manual sampling for successful detection, including the Thermedics Detection EGIS™ explosive detection system (EDS). With a manual system samples are usually obtained from a surface by wiping or vacuuming the surface manually. Although this method of sample acquisition has proven to be effective, it is also labor and time intensive, and potentially intrusive when sampling people. As a result, there is a strong need for an EDS which is geared towards screening high volumes of people. The Thermedics Detection SecurScan$_{TM}$, shown in Figure 1 is designed to fill that need, and is specifically intended for use in airport security checkpoints for the screening of passengers boarding airplanes.

This paper will describe the operation of the SecurScan, present the results of performance evaluations performed in the laboratory, and briefly discuss the upcoming field testing.

Structural Overview

The SecurScan contains six main subdivisions of its components: the archway or sampling module, the analysis or chemistry module, the transport module, the support components module, the internal electronics, and the user interface microcomputer.

1. Sampling Module

The sampling module is composed of the wands which acquire the sample from the passenger, the transport conduit, and the regenerative blowers which provide the impetus for sample collection and transport.

Figure 1 Thermedics Detection SecurScan

The wands are composed of a series of 10 paddles, with each paddle being 10 centimeters wide, 1 meter long, and angled 30° with respect to the horizontal. The wands have thirty-six 1.5 mm diameter holes spaced an inch apart along their length. The 30° slant downwards allows the wand to sample along a 40 centimeter vertical swath of the passengers body as they walk through the archway. The vertical distance between each hole is one half inch. Also, the length of the wands is such that almost every part of the wand overlaps with part of another wand either above or below it. As a result, a sampling hole occurs every 7 mm of vertical height. This translates into direct contact with a sampling hole of nearly 30% of the vertical height of the passenger.

The transport conduit is used to conduct the sample-laden air stream from the archway to the transport module. The danger in using long lengths of transport piping is the potential of sample loss to the walls of the conduit. Any explosives that settle out of the air stream during transport may also become re-entrained during future sampling, resulting in false alarms or an elevated background.

During the proof of principal phase of SecurScan development it was discovered that only the less volatile explosives, which occur as particulates, are likely to settle out of the air stream. Upon conducting an evaluation of sample loss as a function of the speed of the air stream transporting the sample through the conduit, it was found that sample loss decreased dramatically when the air speed exceeded 10 m/s, and that for air speeds exceeding this critical value the sample loss over a 3 meter length of conduit was below the noise level of the experiment. Furthermore, the carry-over rate from sample to sample literally dropped to zero for cases where the signal level for the initial sample was below the saturation level of the detector. A 4 mm diameter bleed hole at the entrance end of the wands provides cross current flow that aids in sample transport. The wands are designed such that a filter can be installed at the ends of the wands to filter the air flowing into the bleed hole, thus reducing the chemical background.

The regenerative blowers provide the vacuum suction that is instrumental in removing the sample from the passenger. The blowers are capable of high volumes of air flow at a large (1/4 atm) pressure differential and are rated for continuous operation. The regenerative blowers are also highly reliable,

Figure 2 Transport Module Assembly

having a MTBF in excess of 10,000 hours with no maintenance requirements. In addition, with a noise level of just over 60 dB, the blowers are relatively quiet.

2. Transport Module

The transport conduit in the sampling module terminates at the transport module. In the transport module the sample is extracted from the air stream, and transported to the chemistry module for analysis. A picture of the transport module assembly is shown in Figure 2. The transport module presently includes a bypass assembly or collection station, a desorb station, and two filter wheels and drives.

Since the vacuuming action generated in the wands necessarily pulls a lot of air into the system with the sample, it is necessary to concentrate the sample by filtering it from the air. The sample collector or filter is composed of a porous Teflon membrane with an open area of about 85% and a pore size of approximately 75 μm (as determined by using a methanol bubble point test). Strips of a chemically absorbent polymer on the collector face provide the mechanism for the collection of the vaporous, more volatile explosives. Although the large amount of open area in the filter permits a high volume of air flow, the small pore size traps particulates in its interstices. However, the impedance of the filter is such that only a fraction of the total air flow needed for effective sampling can be pulled through the sample collector.

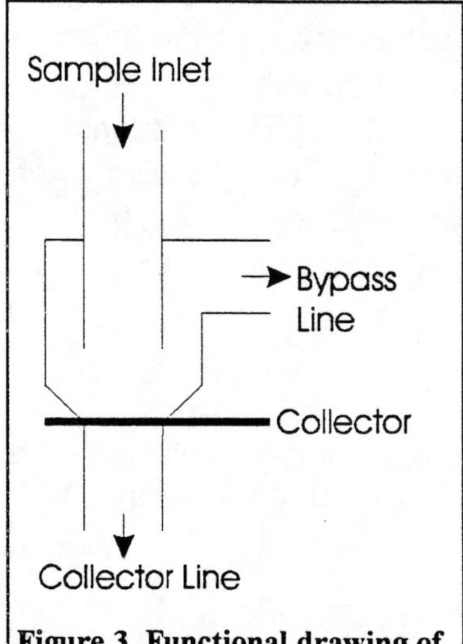

Figure 3 Functional drawing of the bypass assembly.

The geometry of the bypass assembly, shown in figure 3, forces the particulates in the air-stream to impinge upon the filter/collector. Particulates that strike the filter are held in place by the pressure differential across it. This permits the collection of over 90% of the particulates in the air stream, although only 1/3 of the total air flow actually passes through the collector. However, since the momentum of the explosive vapors is roughly equal to the momentum of the air surrounding it, the vaporous explosives will only be collected in proportion with the fraction of the air stream that actually passes through the collector. In practice this is not a deficiency, since the highly volatile explosives produce large quantities of vapor, and a significant amount of the explosives will be available for sampling.

The desorb station is comprised of a heated metal block and the snout, or interface to the chemistry module. The process of transferring the sample from the collector to the chemistry module is called desorbing. The collector is desorbed by sealing it between the heated block and the chemistry module interface. The sample is then flushed with pre-heated purified air into the chemistry module, where it is trapped for injection into the GC system..

The transport module contains two filter wheels. One wheel contains 16 sample collector filters, which are used sequentially as the filter wheel rotates the filters from the collection station, to the desorb station (where the sample is thermally desorbed into the chemistry module), to a cleaning station, and then eventually back to the collection station. Each filter can be used approximately 200 times, for a total of 3200 samples per wheel. If the traffic rate at the security checkpoint where the SecurScan is stationed averages 5 people per minute, each wheel could be used for 10 hours of continuous operation. In a typical security station, it would only be necessary to change the filters once a day. The other wheel contains the snout filters which protect the chemistry module from any particulates that may escape the sample collector during desorption. When the snout filter becomes clogged the wheel rotates, presenting a fresh snout filter.

3. Chemistry Module

The chemistry module is a high speed gas chromatograph connected to a chemiluminescence detector based upon the $NO + O_3 \Rightarrow NO_2 + O_2 + h\nu$ reaction. The SecurScan chemistry module is very similar to the chemistry module used in the Thermedics Detection EGIS$_{TM}$ EDS. Laboratory testing has demonstrated that the false alarm rate, the sensitivity, and the selectivity will also be similar to that of the EGIS. Since a detailed explanation of the chemistry module technology has been given elsewhere[1], only some of the details specific to the SecurScan chemistry module will be presented here.

The chemistry module is composed of two cold spot clusters, a switching valve, two selective elements or chromatographic columns, and three pyrolitic converters. A cold spot cluster consists of multiple cold spots acting together to trap and concentrate the analytes of interest. The first cold spot in the chemistry module is located directly downstream of the desorb station. The explosives that are thermally desorbed off of the sample collector in the transport module flow from the transport module interface, through a switching valve and into the first cold spot cluster. In order to accommodate the high flows needed for fast sample transfer, the desorb flow is split down three parallel lines. Each line contains either one cold spot alone, or two cold spots in series. The first cold spot cluster has a total of five separate cold spots. The cold spots that are downstream of another cold spot, or the secondary cold spots, are actively cooled to below room temperature with thermoelectric peltier coolers. The cold spots that are first in the sample stream, or the primary cold spots, are cooled by air flow only.

After being concentrated in the cold spots, the sample is injected through a split onto two chromatographic columns. Next the chemicals in the sample are separated in the chromatographic columns, and then thermally degraded as they pass through the pyrolytic converter. Any nitric oxide present in the analyte stream following pyrolysis is reacted with ozone in a reaction chamber and detected with an infra-red sensitive photomultiplier tube detector.

4. Support Components Module

The support components module consists of the supporting electronics, the vacuum pump, the carrier gas supply, the ozone generators, and all other components that are essential to the function of the chemistry module, but not directly part of it.

5. Embedded Electronics and Host Computer

The heart of the embedded electronics system is a 486DX-2 66 Mhz computer. This computer and all of its peripherals are mounted into an STD card cage and communicate through an STD-16 bus.

The computer controls all processes which occur in the SecurScan, including temperature control, data acquisition, the monitoring and conversion of analog sensory inputs, digital I/O monitoring and control, and data transfer via a serial communications link.

The embedded system is a stand-alone system which does not require input from the host computer in order to operate. The raw data resulting from the analysis of any samples is downloaded through a serial link from the embedded system to the host computer, which processes the data and displays the results to the user. The host computer performs all user interface functions and data analysis.

Sampling Performance

The development process for the SecurScan included development efforts evaluating different mechanisms for sample acquisition. As a result of this development effort, several sampling innovations have been made which are specific to the SecurScan. The SecurScan is capable of sampling nearly 10 people per minute, or in continuous operation over a 10 hour period 6000 people per day. The sampling system is fully automated, but completely controlled by the passenger as they walk through the portal at their own pace.

One of the tests used to evaluate the start to finish sensitivity performance of the system was the introduction of a controlled source of trace levels of explosives at the wands. This source of explosives consisted of a linen square, or patch, which had been prepared with a quantified level of a target explosive in the sub-microgram range on its surface. The amount of explosive deposited on the patch is typical of the low level of contamination expected in real scenarios. The patches were prepared for our use by the FAA Technical Center. The method used for the preparation of the patches has been detailed in other publications[2]. For security reasons, the exact results of the performance evaluation or any quantification of the sensitivity of the SecurScan cannot be presented here.

The patches were sampled by gently wiping the contaminated surface of the patch over the wand face one time only. A total of six of these pre-prepared patches were sampled in this manner. Of the six patches sampled, all six were detected. From start to finish 4% of the explosives that was known to be on the patch was recovered and detected.

After sampling and detecting a patch a clean-up procedure was followed to characterize the amount of the explosive that remained either in the system or on the outside of the wand where the sample was introduced. The clean-up procedure involved first running a full sample acquisition and analysis cycle without introducing any further contamination or touching the sampling wands. For all six samples, the sampling cycle that immediately followed the sampling of the patch indicated that no residual explosives contamination remained anywhere inside of the system. The next step involved wiping the outside of the sampling wand, where the patch had been introduced, with a clean paper towel while running another sampling cycle. The purpose of this exercise was to determine if any contamination remained on the outside of the wand. In all six cases no contamination remained on the outside of the wand, demonstrating that for this level of contamination, the recovery is instantaneous.

Field Testing

As a prelude to a full field test, laboratory tests were performed to simulate usage of the SecurScan prototype at an airport. The purpose of this evaluation was to 1) debug the operating software and the user interface and 2) evaluate the physical ruggedness of the instrument. This test was accomplished by repeatedly sampling a large number of employees. A total of 1500 samples were collected in six hours spread over three days. No mechanical failures were encountered during the course of the testing. In addition, any user interface glitches that occurred were promptly corrected and re-evaluated before the testing was concluded.

A two week field test of the prototype will be conducted starting in the last week of October. The SecurScan will be installed in the passenger terminal of a major airport after the standard security checkpoint. The instrument will not be part of the checkpoint security and will be positioned in such a way as to not interfere with normal terminal operations. Volunteers from the passenger stream will be solicited to pass through the instrument. During these field trials data will be collected on;

1. What is the time required for the average passenger to transit the sampling system? Is the transit time related to differences in the age, sex, or physical size of the passengers?
2. How efficient is the sampling system in terms of contact between the passenger and the sampling wands? This will be qualitatively evaluated by monitoring the passengers as they move through the system. The passage of the passengers through the system will be videotaped to aid in later evaluation.
3. How accepting are the passengers of the sampling system? The passengers will be questioned after the test do determine whether they find interaction with the SecurScan to be intrusive or benign.
4. What is the average chemical background from the passengers and the airport environment? Are there any compounds in this environment which will interfere with the detection of explosives? It must be realized that this background information will be specific to this airport at this time of year and may change with location and time.
5. What are the nuisance and false alarm rates? A nuisance alarm is caused by persons who have legitimate reasons for being contaminated with explosives, such as someone who uses nitroglycerin as a heart medication. A false alarm is one which cannot be explained by either the presence of or exposure to explosives.
6. How robust is the mechanical portion of the SecurScan. The passengers will be directly interacting with the sampling wands. These wands will be routinely monitored for signs of wear or fatigue.

At the end of these field tests, the data will be evaluated to adjust the system design for optimum efficiency and passenger acceptance.

Acknowledgments

The authors would like to thank the United States Federal Aviation Administration for its support under contract DTFA03-87-C-00003 under which most of the work described in this paper was performed.

A special thanks to Dr. Frank Fox of the USFAA Technical Center for supplying the patches used in the performance evaluation of SecurScan.

---

[1] Rounbehler, D; MacDonald, S.; Lieb, D.; and Fine, D. (1991) "Analysis of Explosives Using High Speed Chromatography with Chemiluminescent Detection, " Proceedings of the First International Symposium on Explosive Detection Technology, November 13-15, 1991, 703-713.

[2] Fox, Sisk, DiBartolo, Green, and Miller, "Preparation and Characterization of Plastic Explosives Suspensions as Analogs of Fingerprint Derived Contaminants for Use in Certification of Trace Explosives Detection Systems", Proceedings of the Third Workshop of the ICAO Ad Hoc Group of Specialists on the Detection of Explosives, FAA, William J. Hughes Technical Center, Aviation Security Laboratory, Atlantic City International Airport, NJ, Oct. 1995.

SESSION 3

Detection of Concealed Contraband

Screening technologies for detection of swallowed packages of narcotics

Lowell J. Burnett, Erik E. Magnuson, Alan G. Sheldon, and S. Kumar

Quantum Magnetics, Inc.
7740 Kenamar Court, San Diego, CA 92121
Phone: (619) 566-9200; Fax: (619) 566-9388

ABSTRACT

An increasingly popular method of transporting modest quantities of narcotics across international borders is to employ "swallowers." These are people who typically enter the country as international airline passengers after swallowing small, water-tight packages of heroin and/or cocaine.

Rapid and accurate identification of swallowers in the airport environment poses difficult technical challenges. Commonly used medical inspection technologies fall into one of two categories. Either they are unsuitable for widespread use, or they do not provide adequate information. An example of the former is X-ray scanning, while an example of the latter is ultrasonic imaging.

Quantum Magnetics has developed a system to screen selected airline passengers for the presence of swallowed narcotics. The system utilizes magnetic resonance (MR), which provides the physical basis for the magnetic resonance imaging (MRI) systems widely used in the medical community as an alternative to X-rays. The system is currently operational, and laboratory performance testing is complete. Both the design of the system and its performance will be discussed.

This work was sponsored in part by the Office of National Drug Control Policy and the U.S. Customs Service.

Keywords: Cocaine, heroin, drug smuggling, magnetic resonance, non-invasive detection, aviation security.

1. INTRODUCTION

According to a recent article in The Economist, fully one-third of the women in British jails are incarcerated for the transportation of swallowed narcotics into the UK. And during a recent four-month period, approximately 300 "swallowers" were arrested in the Saudi Arabian city of Jedda. And on average, one swallower a day is arrested in Miami. Clearly, this is a worldwide problem.

The identification of swallowers, by direct detection of the swallowed narcotics, is a problem involving both technical and legal complexities. Quantum Magnetics has solved the technical aspects of this problem by developing a nuclear magnetic resonance (NMR) scanning system that distinguishes between the ingested narcotics and the normal contents of the human body. This system, termed the SDS-1 Swallower Detection System, is quick and accurate, and does not subject the individual to harmful ionizing radiation. The SDS-1 meets the US Food and Drug (FDA) guidelines for medical device safety.

2. PROBLEM DEFINITION

For travellers entering the United States, the average number of narcotics packages ingested by a swallower is 86, and common package weights range from three to 7.5 grams. The average package weight is five grams, and a typical size is 1.5 cm diameter by five cm in length. The average swallower brings in 430 grams (about one pound) of narcotics, although swallowers carrying in excess of one kilogram are not uncommon.

After ingestion, after travel to the airport, after waiting in the airport, and after sitting on the airplane, for a cumulative total of many hours, the narcotics packages reside in well defined places in the human digestive system. Specifically, many packages end up primarily in the ascending colon, the descending colon, and the transverse colon. The SDS-1 screening system is designed to interrogate these regions.

3. NUCLEAR MAGNETIC RESONANCE

Nuclear magnetic resonance (NMR) arises from the magnetic properties of atomic nuclei [1]. When immersed in a steady magnetic field, B_O, the nuclei align with the field and precess at a characteristic frequency called the NMR resonance frequency which is proportional to the strength of the applied magnetic field, B_O. For hydrogen nuclei, or protons, a magnetic field of one tesla (10,000 gauss) produces an NMR frequency of 42.6 MHz.

An externally applied pulse of radio-frequency (RF) field at the NMR resonance frequency causes the nuclei to tip away from their equilibrium position and, while still precessing, induce an NMR signal into an adjacent detection coil. In NMR jargon, the signal is often called the free induction decay (FID), and the detection coil is often called the sample coil.

The low-power NMR electronics of the SDS-1 are housed on four circuit boards that plug into the ISA bus of an IBM-compatible PC [2]. Timing functions are provided by the pulse programmer board. RF pulses at the NMR frequency are formed on the RF board, then amplified and applied to the sample coil through a multiplexer circuit. The NMR signal from the inspection region is amplified by a low noise preamplifier, then routed back to the RF board for further amplification and detection. The detected output is converted into digital form by the A/D Board, then stored and processed. A fourth board, the control board, monitors the state of the system and adjusts the current in the magnet coils to produce the desired magnetic field.

4. SYSTEM DESCRIPTION

A photograph of the SDS-1 Swallower Detection System is shown in Figure 1. The SDS-1 consists of a large low-field electromagnet, a detection coil that encloses the lower torso on three sides, an RF amplifier, and the computer-housed NMR electronics package. The suspect stands in the magnetic field, is properly positioned with a small hydraulic lift, then subjected to a series of high intensity, low duty cycle RF pulses. These RF pulses produce an NMR response from the hydrogen nuclei within the suspect.

Figure 1. The SDS-1 Swallower Detection System.

In operation, the SDS-1 utilizes an RF pulse sequence that produces distinct signals from the crystalline solid narcotics and the liquid and semiliquid components of the human body. These NMR signal components are readily separable and a positive reading, above a defined threshold, for the crystalline solid component indicates the presence of swallowed narcotics. In this way, the effects of person-to-person variations in the NMR response are minimized.

The SDS-1 requires a minimum floor area 4.3 meters on a side, with a minimum clear height of two meters. The system weighs approximately 5 megagrams and requires an 80 ampere, 208 volt, three-phase line for operation. The SDS-1 dissipates about 10 kilowatts of heat into the room, and the temperature control system at the site must be able to accommodate this extra load.

5. SYSTEM PERFORMANCE

A series of laboratory tests using a phantom to simulate the human torso, in which controlled amounts of narcotics and simulants were placed, was used to determine the detection sensitivity of the SDS-1 Swallower Detection System. Since the NMR signal from the narcotics results from a subtraction process in the time domain, the false alarm rate and the probability of detection are determined by the overlap of two Gaussian functions (one corresponding to the system response with no narcotics present and the other to the response with narcotics present), the system sensitivity, and the choice of threshold voltage.

When set for a 2.0% false alarm rate with no narcotics present, the laboratory tests revealed that the SDS-1 is capable of detecting 30 grams of ingested cocaine in a 30 second scan with a probability of detection of 92.7%. Similar results were obtained for heroin and for narcotic simulants which, like the cocaine, were also located in a dense liquid and semi-liquid matrix within the phantom. Naturally, larger quantities of narcotics produce higher detection probabilities. In addition, several straightforward improvements are expected to improve the current SDS-1 performance substantially.

The SDS-1 Swallower Detection System is fully automated, and only a single operator is required. After the suspect is positioned, screening begins at the touch of a button and takes approximately 30 seconds. The results are displayed as a "Pass/No Pass" response, and no special operator training or image interpretation is required. Furthermore, the SDS-1 system is safe. It does not emit ionizing radiation, and falls within the FDA guidelines for magnetic field exposure by at least a factor of ten.

6. ACKNOWLEDGMENTS

The authors would like to thank Dr. Victor Pocaro and Mr. Douglas Smith of the U.S. Customs Service for their continuing interest, support, and oversight of this complex and multifaceted development program. We would also like to thank Dr. David Shykind of Intel Corporation for his many contributions to the technology.

7. REFERENCES

1. A. Abragam, *The Principles of Nuclear Magnetism*, Clarendon Press, Oxford (1961).

2. L.J. Burnett, "Liquid Explosives Detection," Presented at the EOS/SPIE International Symposium on Substance Identification Technologies, Innsbruck, Austria, October 1993. Published in the Proceedings of the SPIE **2092**, 208 (1993).

Initial test and evaluation of the millimeter-wave holographic surveillance system

D. L. McMakin, D. M. Sheen, A. Schur, W. M. Harris, and G. F. Piepel.
Pacific Northwest National Laboratory*
P.O. Box 999
Richland, WA, USA

ABSTRACT

A test and evaluation pilot study was conducted in January 1996 at Sea-Tac International Airport in Seattle, Washington to determine the initial effectiveness of the Millimeter-wave Holographic Weapons Surveillance System. This is a new personnel surveillance system for the detection of concealed metal, plastic, and ceramic weapons and other threatening materials. Two different frequency bands were used in the study: Ku band (12 - 18 GHz) and Ka band (27 - 33 GHz). Over 7000 Millimeter-wave (MM-wave) holographic images were obtained on 21 different models (10 male and 11 female). The 7000 images were used to produce simulated real-time surveillance system videos. The videos were constructed by obtaining 36 images of the models at 10 degree increments for 360 degree coverage. A library of two hundred videos were produced for this pilot study: 100 at Ku band and 100 at Ka band. The videos contained either a threat (firearm, knife or explosive) or no threat. The threats were concealed at different locations on the models. Various innocuous items and different clothing combinations were also used in the construction of these videos. Twenty-nine certified Sea-Tac screeners were used in the initial test and evaluation of this new surveillance technology. Each screener viewed 160 MM-wave videos: 80 Ku band and 80 Ka band. The ratio of non-threat to threat videos per band was three to one. Test and evaluation software was developed to collect data from the screeners online for the type and location of threat detected. The primary measures of screener performance used to evaluate this new technology included, the probability of detection, the probability of a false alarm, measures of screener sensitivity and bias, and threat detection time.

Keywords: Explosive detection, weapons detection

1. INTRODUCTION

The Millimeter-Wave Holographic Weapons Surveillance System (MMW-HWSS) is a new technology under development at the Pacific Northwest National Laboratory that has demonstrated great promise for the detection and identification of metal, plastic, and ceramic weapons and threatening materials concealed on airline passengers[1,2]. Millimeter waves readily penetrate clothing barriers and the holographic imaging technique forms very high resolution images of the concealed threats. This new technology is at a stage of development where airport screener performance baseline studies are necessary to determine the preliminary effectiveness of the technology for detecting concealed weapons and other threatening material.

To evaluate the MMW-HWSS, a pilot study was conducted to assess the performance of security screeners with the system. Security screeners were asked to view videos of passenger models who were possibly concealing threats and to indicate whether the models had a threat object. The security screeners were also asked to identify the type and locations of perceived threats. Because the videos generated by the MMW-HWSS can be either high-resolution images (Ka band) or low-resolution images (Ku band), both image types were presented to the security screeners to determine if resolution affects detection performance.

The data obtained from the pilot study was statistically analyzed for individual screeners, all screeners combined, individual videos, and all videos combined. These statistical analyses were used to: 1) determine whether certified airport screeners could detect concealed weapons and explosive threats using the MMW-HWSS at both the Ku band and the Ka band, and 2) investigate whether the screener/system performance depends on various factors such as the type of threat, threat placement on the body, innocuous items, passenger gender, and passenger clothing. Additionally, a statistical analysis was performed on

---

* PNNL is a multiprogram laboratory operated for the U. S. Department of Energy under Contract DE-AC06-76RLO 1830 by Battelle Memorial Institute

screener decision time to establish a baseline for this new surveillance technology. Lastly, each screener completed a debriefing after completing the detection task to solicit their opinions on this new surveillance technology and how the system might be used at airports, how they used the system during the test, improvements needed by the system, and their overall opinion of the MMW-HWSS as an aid to threat detection. The results of these analyses are summarized in the following discussion.

2. TEST SYSTEM DESCRIPTION

The MMW-HWSS obtains an image by directly measuring the phase and amplitude of MM-wave wavefronts scattered from concealed objects. A wavefront reconstruction process is then applied to the phase and amplitude data to form an image on the computer screen in 64 levels of gray scale. The system shown in Figure 1 was used to generate holographic images of passenger models carrying concealed innocuous items and real or simulated threats.

The MMW-HWSS uses longer wavelengths than optical systems to produce images of potential threats. The use of these longer wavelengths produces threat images that are of noticeably lower resolution (less sharp) than a similar X-ray image or other optical image. The longer wavelengths do, however, penetrate many optically opaque materials, such as clothing, which permits the detection of threats that are otherwise hidden from the screener.

Figure 1. Current Wideband Holographic Millimeter-Wave Imaging System

2.1. Ku Band and Ka Band Holographic Arrays

Two wideband holographic arrays were used to take MM-wave scans of passenger models: a lower-frequency Ku band array and a higher-frequency Ka band array. Both system arrays are capable of scanning one view of a person in 1 second.

- The Ku band (12 - 18 GHz) array consists of 74 elements. The lateral resolution of this system is 1 cm. (Lateral resolution refers to resolution in the image plane, as opposed to depth resolution).

- The Ka band (27 - 33 GHz) array consists of 128 elements. The Ka band lateral resolution is 0.5 cm, which is twice as good as the Ku band images. The better lateral resolution should help the screener to detect concealed weapons and discriminate them from innocuous items. The Ka band system, however, is more expensive than the Ku band system.

2.2. Construction and Selection of MM-Wave Videos

Selecting the number and type of MM-wave videos was the initial part of the experimental design. The library of videos needed to be representative of passengers, threats, and threat placements that might occur in an airport screening situation. The large number of possible levels of these factors--and the correspondingly large number of combinations of levels of these factors--made it impractical to construct an exhaustive collection of videos. Thus, representative levels of each factor were chosen:

1) simulated passengers -- 10 male passenger models and 10 female passenger models of various body types (11 female models were used to obtain MM-wave videos, but only 10 were shown to the security screener during the test), each wearing one of three categories of clothing and carrying two of nine types of innocuous items

2) threats -- three kinds of firearms, two kinds of knives, and two kinds of plastic explosives

3) threat placements -- six different locations where the threat might be concealed on a passenger's body.

Passenger models were scanned to create pre-recorded MM-wave videos for use in the subsequent screener testing. For each video, 36 gray-scale images or frames of the person under surveillance were obtained by scanning at 10-degree increments for full 360-degree coverage around the body axis. These 36 frames were then integrated to produce a MM-wave holographic video of the passenger model. Figure 2 shows a typical MM-wave video sequence.

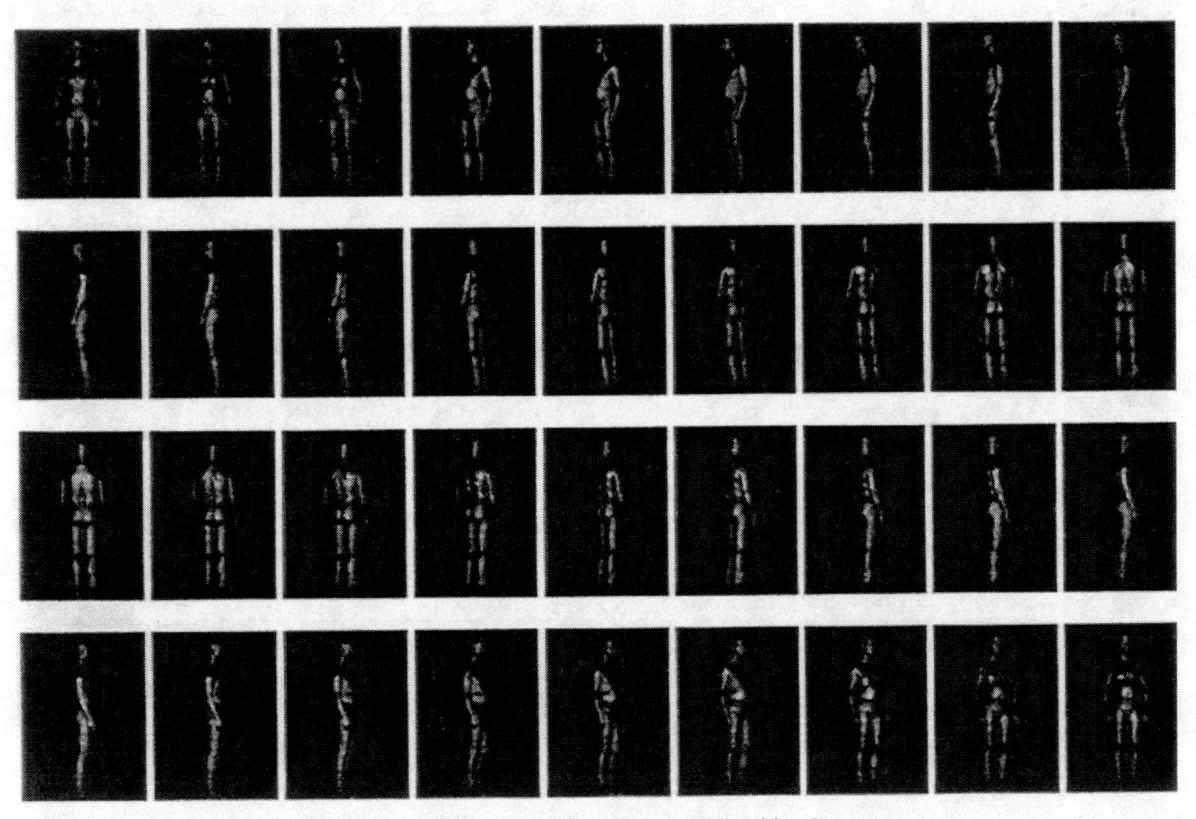

Figure 2. Typical MM-Wave Holographic Video Sequence

The MM-wave videos were displayed by computer to simulate a real-time surveillance system. The simulated motion can help an airport screener differentiate between objects that are concealed on the body by showing multiple views. A selected frame rate of 4 frames per second resulted in one complete body rotation every 9 seconds.

The possible threats and threat placements were combined in a systematic way to yield 40 threat/placement combinations. Then models, clothing, and innocuous items were randomly assigned to the 40 combinations in a structured way so as to maintain a balance of models, clothing, and innocuous items.

A natural pairing experimental design framework was used to ensure that task difficulty was equated across the two bandwidths (Ku and Ka) and between threat and innocuous videos. A test library of 160 videos was constructed to contain 80 Ku band videos matched in content with 80 Ka band videos. For each band, the 80 videos consisted of 40 threat videos and 40 innocuous videos, where each threat video had a corresponding innocuous video with the same model, clothes, and innocuous items. The two natural pairings (Ka versus Ku, and threat versus innocuous) eliminate other possibly confounding factors that could complicate data analyses.

Table 1 shows the Threat/Placement Matrix used in the construction of the MMW-HWSS videos. The table shows that six threat placements and seven threats were used in the construction of the MM-wave videos for a total of forty-two combinations of threat and threat placements. Four of these threat configurations are not likely to appear (such as a large weapon inside the thigh) and therefore were not considered. However, two of the threat configurations may differ in detectability, depending on the sex of the model. In these cases both a male and a female model were used with these two threat configurations (in 4 threat videos). For the remaining 36 threat configurations, the sex of the model was not considered of great importance. Therefore, a total of forty threat configurations were required for a complete set of threat videos. Each model appeared in two of three possible types of clothing: informal wear, layered clothing, or formal clothing (business suits). Each model carried one of 36 possible combinations of two out of nine innocuous items. The nine innocuous items used in this pilot study included a cigarette pack, cigarette lighter, calculator, keys, coins, ink pens, glasses case, wallet, and check book.

Table 1. Threat/Placement Matrix for MMW-HWSS Videos (in either Ku or Ka Band)

PLACEMENT	SIDEARM			KNIFE		PLASTIC EXPLOSIVE	
	Glock	Raven Arms	Flare Gun	Fixed Blade	Folding Blade	Bulk	Sheet
Underarm	S_1	S_2	S_3	S_4	S_5	S_6	S_7
Chest	S_8	S_9	S_{10}	S_{11}	S_{12}	S_{13}	S_{14}
Small of the Back	S_{15}	M_1, F_1	S_{16}	S_{17}	M_2, F_2	S_{18}	S_{19}
Hip/Thigh	XXX	S_{20}	XXX	S_{21}	S_{22}	S_{23}	S_{24}
Inside Thigh	XXX	S_{25}	XXX	S_{26}	S_{27}	S_{28}	S_{29}
Calf/Ankle	S_{30}	S_{31}	S_{32}	S_{33}	S_{34}	S_{35}	S_{36}

XXX - **Four (4) threat configurations that are not likely to appear -- Therefore they were not considered.**

M_i, F_i - **Two (2) threat configurations that may differ in detectability, depending on the sex of the model -- Both a male and a female model were used with these threat configurations.**

S_i - **Thirty-six (36) threat configurations for which the sex of the model is not considered of great importance**

Figure 3 shows four of the 36 Ka band images used in the production of a threat video. In this configuration, the model has a concealed Glock in the small of the back. The model is wearing informal clothing (sweater and jeans) and has a cigarette pack in her back pants pocket and keys in her left front pants pocket.

Two sets of threat videos (each corresponding to the threat/placement matrix in Table 1) were required: one for Ku band videos, and the other Ka band videos. Similarly, two sets of innocuous videos at Ku band and Ka band were required. The conditions for threat and innocuous videos at both bands were identical. That is, the model, clothes, and two innocuous items used for a given threat video were also used in a corresponding innocuous video. For a given frequency band, a threat video and its corresponding innocuous video are referred to as a *video pair*.

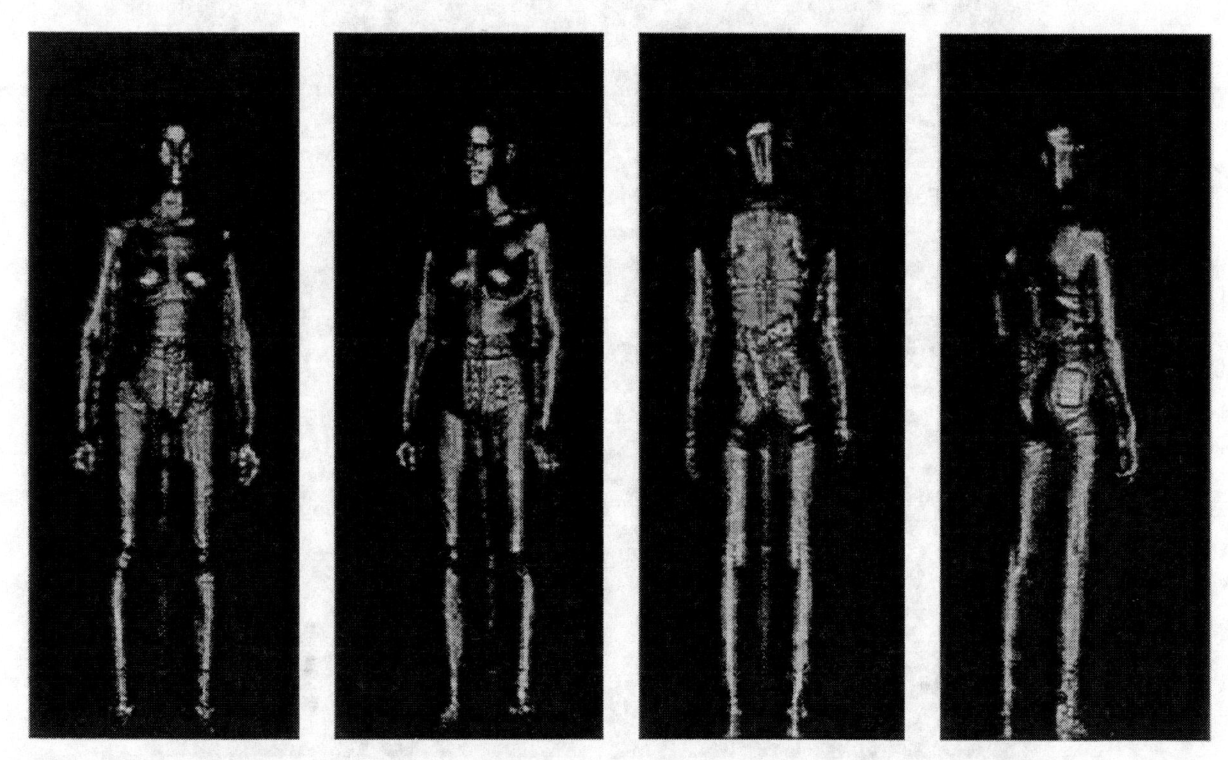

Figure 3. Four of 36 Ka band images used in the production of a typical threat video

Figure 4a shows four of 36 Ka band images used in the production of an innocuous video and Figure 4b shows the threat video images that are identical to the model configuration in Figure 4a except that simulated sheet explosive has been strapped around the outside of the model's left thigh. In this configuration, the male model is wearing informal clothing (jeans and a flannel shirt) and has a calculator in his left front shirt pocket and keys in his front right pants pocket.

2.3. Systems Used to Implement the Testing Interface

Four identical 133-MHz Intel Pentium PCs with 21-inch, high-resolution, flat-screen color monitors were used to display the MM-wave videos and record screener responses in the test. All machines used a 64-bit accelerated graphics controller with 2MB WRAM (Windows Random Access Memory).

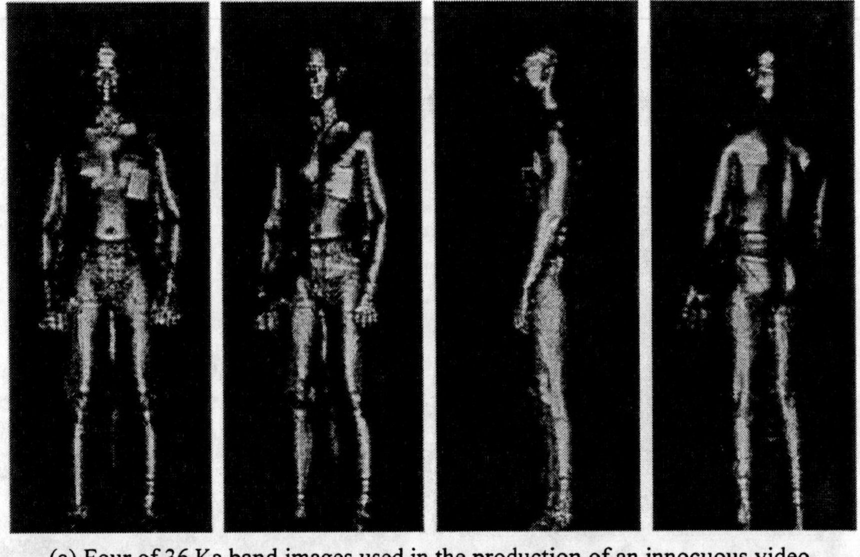

(a) Four of 36 Ka band images used in the production of an innocuous video

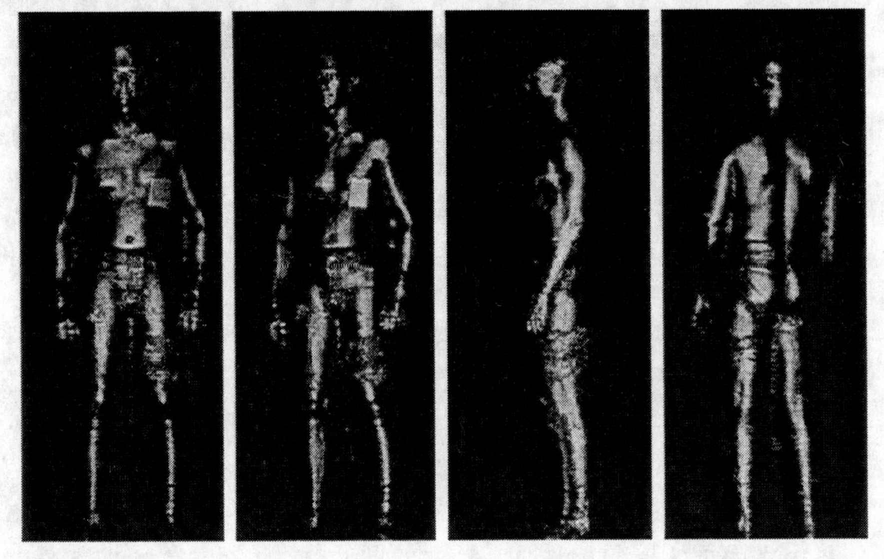

(b) Four of 36 Ka band images used in the production of a threat video

Figure 4. Innocuous and threat *video pair* images of a male model with concealed sheet explosive on left thigh. a) innocuous video images, b) threat video images.

Macromind Director 4.0 multimedia software was used to develop the user testing interface,. Figure 5a shows the screen used for displaying MM-wave videos and registering a screener's response. The screener indicated either the absence or presence of a detected threat by using the computer's mouse to click on either the "No" or "Yes" button.

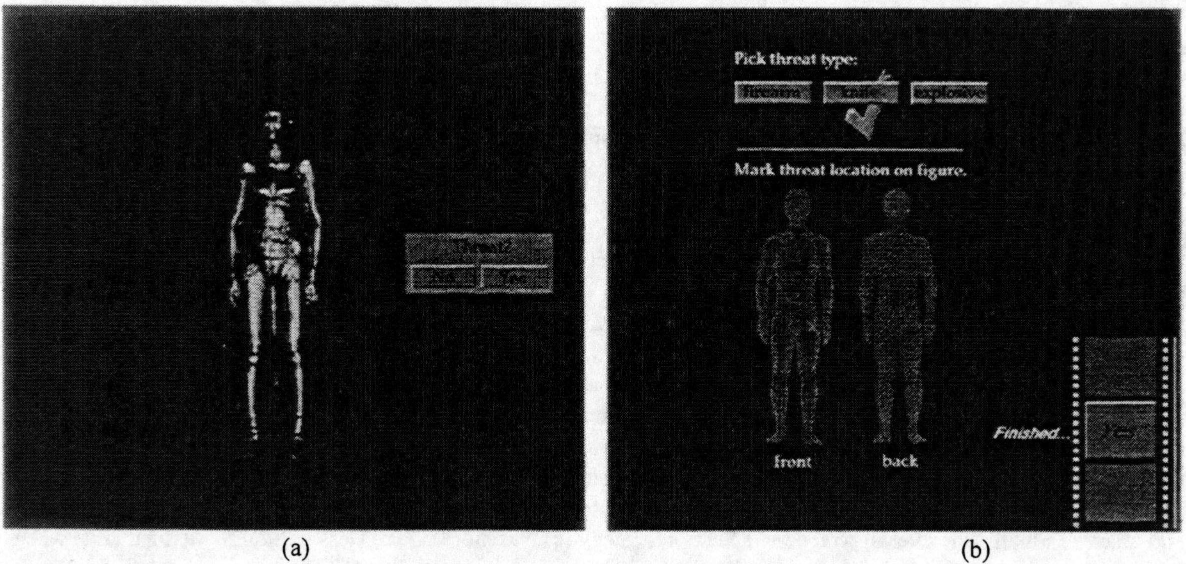

(a) (b)

Figure 5. Test interface screens used to capture screeners response on-line. a) screen for image display and threat detection, b) screen for indicating threat type and location..

Figure 5b shows the screen for indicating the type and location of detected threat. Screeners indicated threat location by pointing and clicking the mouse on a front or back depiction of a human form. The software recorded the threat location data by assigning the screener's response to one of 39 body zones depicted in Figure 6.

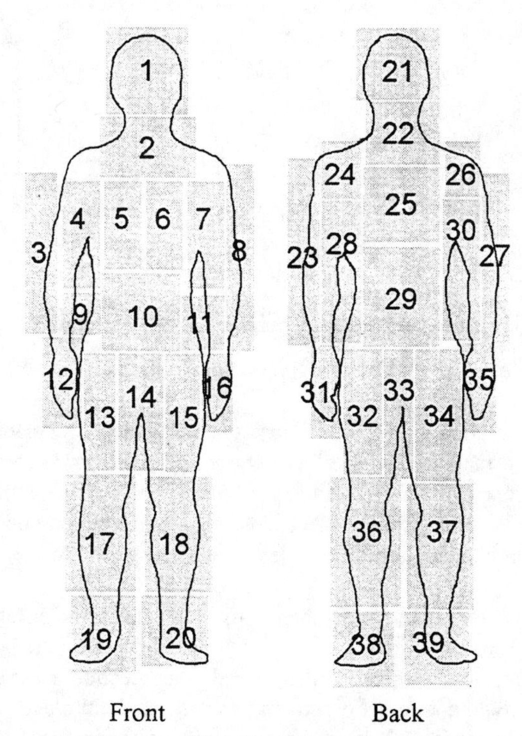

Figure 6. Software Assignment of Screener-Identified Threat Locations to Body Zones (1-39)

3.0. DISCUSSION OF TEST RESULTS

This section discusses the pilot study screener performance results for the new weapons detection surveillance system. The primary measures of screener performance used to evaluate this new technology included the probability of detection, the probability of a false alarm, measures of screener sensitivity and bias, and the threat detection time. Measures of screener bias and sensitivity were used because of their widespread and successful use in Signal Detection Theory[3,4,5].

3.1. Summary of Results Across Screener

Figure 7 shows that certified airport screeners successfully detected concealed threats (guns, knives and plastic explosives) on people by observing MM-wave videos made at both Ku band (12 - 18 GHz) and Ka band (27 - 33 GHz). The average probability of detection (P_d) over all screeners and videos for Ku band was 82.24% and for Ka band was 72.57%. Screeners apparently could differentiate between concealed threats (signals and noise) and noise in the MM-wave videos with adequate effectiveness. There was no statistically significant difference between P_{fa} values for Ku band (34.48%) and Ka band (30.98%). These false alarm statistics indicate that screeners were moderately successful at discriminating between threats and innocuous images for both frequency bands. There was a statistically significant difference in the sensitivity between the two bands as indicated by d'. The data show that the difference in the mean d' values of the N and S+N distribution was larger for the Ku band (1.48) than the Ka band (1.22). These data suggest that the Ka band images may have been "noisier" than the Ku band images. Finally, the overall screeners' detection strategy (c) was more liberal (less willing to miss a threat) when observing a threat displayed with the Ku band (-0.313) compared to the Ka band (-0.064). The average screener bias at Ka band is statistically non-significant, indicating that screeners were (on average) essentially unbiased in their detection strategy with Ka band images.

Probability of Detection and Probability of False Alarm

100
80
60
40
20
0
82.24
34.48
72.57
30.98
Pd
Pfa
Ku Band
Ka Band

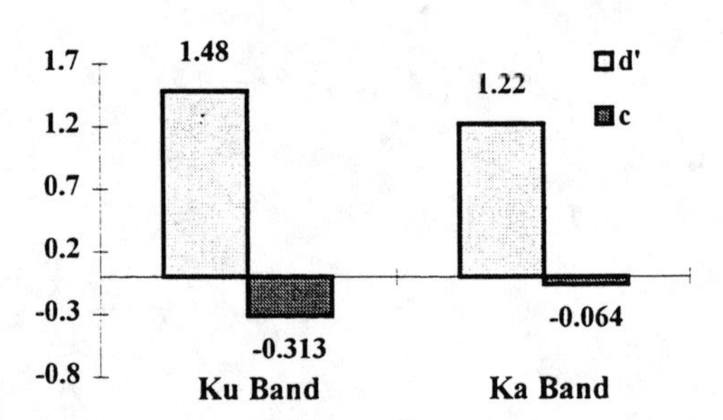

(a) Probability of detection and false alarm by frequency band (b) Screener sensitivity and bias at the two different bands

Figure 7. The Result of the Test and Evaluation By Band and Over All Screeners

It was hypothesized that higher-resolution images would lead to better detection performance across the screeners. The data collected in the study did not support this fundamental hypothesis. The performance measures P_d and d' indicated that screener threat detection performance was better with images produced using the Ku band. There was no statistically significant difference between the P_{fa} values for the two bands. However, the screener bias (c) was more liberal for Ku band than for Ka band, which may explain the higher P_d for videos made using the lower frequency.

A shift in decision bias was observed between the two bands, as indicated by the c statistics. Debriefing interviews suggest that the screeners looked for shapes or conditions that would not be expected on the body. Where the detail or contrast was high, the screener could see the shape of the threat, which served as a detection criterion. In the case of the Ku band images, the shapes of many of the threats could not be easily recognized because the threat detail was beyond the resolution of the MMW-HWSS. The screener, however, could discern that something in the image was not consistent with what might be expected on a person that carried no threats. These anomalies were more often called threats with the Ku band than the Ka band, as indicated by the c statistic and by a high false alarm rate. This more liberal decision criterion with the Ku band may have led to the observed better detection performance.

Data analysis was also performed on results across videos. The summary results across videos are essentially statistically the same as the summary results across screeners shown in Figure 7. Minor differences occur in the resulting averages because of the slightly unbalanced nature of the data, but these differences are not of practical significance.

Figure 8 shows the mean value of P_d for each threat averaged over threat placements, and Figure 9 shows the mean values for each threat placement averaged over threats. These results were obtained from the data analysis by video.

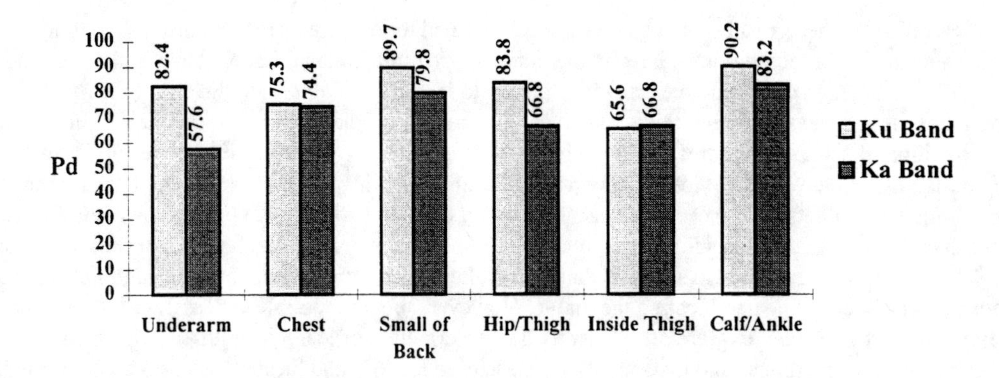

Figure 8. Mean Value of P_d for Threat Placements Averaged Over Threats

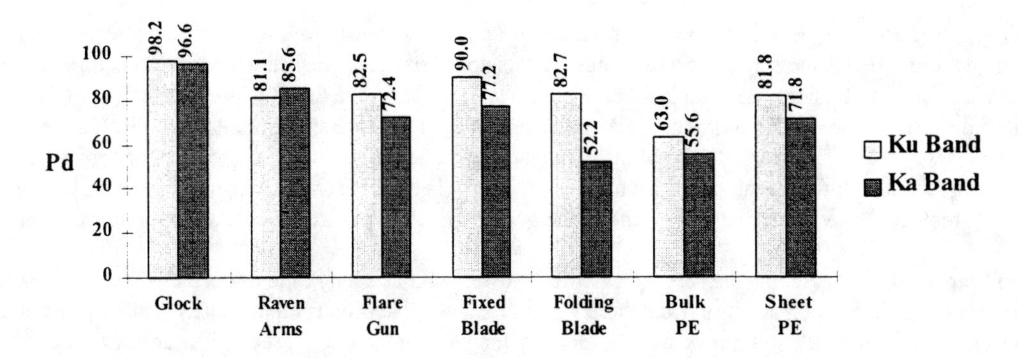

Figure 9. Mean Value of P_d for Threats Averaged Over Threat Placement

Caution must be exercised in viewing the results in Figures 8 and 9 because there are indications the factors "threats" and "threat placements" interact with each other. Averaging over one factor to assess the effects of the other can be misleading when the two factors interact. With this caution in mind, Figure 8 shows that threats on the calf/ankle and in the small of the back were, on average, easiest to detect (largest P_d values considering both Ku and Ka bands). Threats underarm for Ka band videos, and threats on the inside thigh for Ku and Ka band videos, were the hardest to detect, on average. Figure 9 shows that the Glock firearm was the easiest to detect (largest average P_d value for both Ku and Ka bands) and bulk plastic explosive was the hardest threat to detect (smallest P_d value considering both Ku and Ka bands).

It was of interest to assess whether the gender of a model, the type of clothes worn by a model, or the innocuous items carried by a model had any effects on the results. The data provide a limited ability to assess such effects for both threat and innocuous videos. Not all combinations of model gender, clothing type, and innocuous items were investigated, and combinations of these factors were not performed with all possible combinations of threats and placements. Rather, levels of model gender, clothing type, and innocuous items were assigned to the 40 videos (same for threat and innocuous versions of each video) in a structurally randomized manner. It was recognized in the experimental design phase that the structured randomization of these factors provides limited ability to assess the effects of model gender, clothing type, and innocuous items. However, this was deemed acceptable: 1) to keep the pilot study of reasonable size, and 2) because the effects of such factors, if any, were expected to be small. Keeping this caution in mind, the results suggest that screener performance was not noticeably affected by the gender of the passenger. However, the results indicate that different clothing ensembles and various innocuous items may affect the probability of detection and probability of false alarms.

3.2. Summary of Results for Identification of Threat Types and Placements

In addition to determining whether screeners could detect a concealed threat, screeners were also asked to identify the type of threats and their location. This activity went beyond the initial scope of operational issues stated in the Test and Evaluation Plan (TEP) but was included to determine the screener's ability to identify the concealed threat after it was detected. Two screener experience factors may have generally affected the threat type and placement identification results: 1) limited training on recognition of actual threats and 2) inexperience with using a computer mouse. The results show that the screeners performed moderately in threat identification and placement studies for both Ku and Ka bands. The probability of correctly identifying both threats and placements was 50% for Ku band and 55% for Ka band. The result for Ka band is slightly higher than for Ku band, which suggest that increased resolution may improve the tasks of identifying threat types and locations. However, these results also suggests that the resolution for both Ku band and Ka band images/videos may be inadequate for properly identifying and locating the threat. However, another possible explanation for this moderate performance may be that the screeners were insufficiently trained/oriented for identifying threat types and locations. As already discussed above, some threats and placements were easier to identify and locate, such as the Glock handgun in the small the back and on the ankle. Other threat and locations were more difficult to identify, such as the folding blade knife and on the inner thigh and under the arm.

3.3. Summary of Results for Decision Time

During testing, screeners were allowed to view each video for up to 27 seconds. At the end of this maximum viewing period, the video image was removed and screeners had to make a decision about the presence or absence of a threat. The time from the start of a video until a decision was made about the presence or absence of a threat was defined as the "time to decision" or "decision time." The video rotation time was fixed at 9 seconds per rotation and limited to three rotations per video. Because threats were placed on different sides of the body and the model started in the same face-forward position in each video, the first appearance of the threat varied from video to video. Therefore some variation in the decision times for a particular threat may be caused by threat placement on the body and not by real decision time differences.

The decision time results indicate that the screeners in this test were taking a long time to decide when a video contained a threat object, about 19 seconds on average when there was a threat and 24 seconds on average when no threat was present or the threat was missed. These decision times are considerably longer than the 6 to 10 seconds necessary for a future deployed system. Two reasons may account for the long response times:

- Liberal detection strategy --Screeners more willing to have a false alarm than a miss would tend to be more thorough.
- Test environment – Screeners were under no external pressure from passengers or administrators to decide rapidly.

There are no statistical differences between the Ku band and Ka band decision time results for any of the four decision outcomes (correct detection, correct rejection, false alarm, and miss) across the 29 screeners. But for either band, decision times for correct detections (hits) are statistically significantly smaller than decision times for the other three outcomes (correct rejection, false alarm, and miss) Table 2 shows the mean decision times by band and decision outcome.

Table 2. Summary Statistics of Mean Decision Times by Band and Decision Outcome

Outcome	Mean Decision Times by Frequency Band (seconds)	
	Ka Band	**Ku Band**
Correct Rejection	23.40	24.17
False Alarm	24.10	23.88
Detection	18.79	19.46
Miss	23.74	24.33

4.0. CONCLUSIONS

The initial pilot study for the test and evaluation of the MMW-HWSS was described. The primary measures of screener performance - probability of detection, probability of a false alarm, threat type identified, measures of screener response bias and sensitivity, and threat detection time - were used to determine the effectiveness of the MMW-HWSS threat detection at two different frequency bands (Ku and Ka band). The results show that screeners were capable of detecting concealed threats with both frequency bands, although there was some difficulty in distinguishing between threat and innocuous (signal versus noise) cases. Results are inconclusive on which frequency band is superior for concealed threat detection. The results show that the lower frequency (Ku band) had a higher probability of detection (82%) than the higher frequency system (73%) over all screeners. However, the screener bias (c) was more liberal (more willing to have false alarms than misses) for the lower resolution videos (Ku band) than for the higher resolution videos (Ka Band), which may explain the higher probability of detection for videos made using the lower frequency.

The results of the pilot study also show that the threat type and placement affected the screener performance. The screener could readily detect the Glock handgun over all threat locations with both frequency bands, but they had difficulty detecting bulk plastic explosives and threats inside the thigh at both frequency bands.

The screeners' average decision time for detection was significantly lower at both frequency bands than average decision times for correct rejection, misses, and false alarms. The average decision times for detection were 18.8 seconds for Ka band and 19.5 seconds for Ku band. The average decision times for the other three outcomes ranged from 23.4 seconds to 24.3 seconds for both frequency bands. Screeners used an average of about two image rotations to make a detection decision for each frequency band, and about two and two-thirds rotations to make the other three outcome decisions for each frequency band. Screener decision times were affected because the video rotation speed was fixed at 9 seconds per rotation and screeners were not given control over the direction, speed, and progress of the videos (fast forward, fast reverse, pause).

5.0. ACKNOWLEDGEMENTS

The authors gratefully acknowledge the full support of this work by the Federal Aviation Administration (FAA).

6.0. REFERENCES

[1] McMakin, DL, DM Sheen, HD Collins, TE Hall, and RH Severtsen. 1995. "Wideband, millimeter-wave, holographic surveillance systems." EUROPTO International Symposium on Law Enforcement Technologies: Identification Technologies and Traffic Safety, Munich, FRG, SPIE Vol 2511, pp. 131 - 141. International Society for Optical Engineering.

[2] McMakin, DL, DM Sheen, HD Collins, TE Hall, and RR Smith. 1993. "Millimeter-wave, high-resolution, holographic surveillance system." EUROPTO International Symposium on Substance Identification Technologies, Innsbruck, Austria, SPIE Vol 2092, pp. 525 - 535. International Society for Optical Engineering.

[3] Coren, S, and LM Ward. 1989. *Sensation and perception.* Third Edition. Harcourt Brace Jovanovich, San Diego.

[4] See, JE, JS Warm, WN Dember, and SR Howe. November 1994. "Vigilance and signal detection theory: an evaluation of response bias measures." Presented at a poster session of the Psychonomic Society, St. Louis, Missouri.

[5] Wickens, CD. 1992 . *Engineering psychology and human performance*. Second Edition, Harper Collins Publishers, New York.

Detection of contraband concealed on the body using x-ray imaging

Gerald J. Smith

American Science and Engineering, Inc.
829 Middlesex Turnpike
Billerica, Ma. 01821

ABSTRACT

In an effort to avoid detection, smugglers and terrorists are increasingly using the body as a vehicle for transporting illicit drugs, weapons, and explosives. This trend illustrates the natural tendency of traffickers to seek the path of least resistance, as improved interdiction technology and operational effectiveness have been brought to bear on other trafficking avenues such as luggage, cargo, and parcels. In response, improved technology for human inspection is being developed using a variety of techniques. AS&E®'s BodySearch™ X-ray Inspection System uses backscatter x-ray imaging of the human body to quickly, safely, and effectively screen for drugs, weapons, and explosives concealed on the body.

This paper reviews the law enforcement and social issues involved in human inspections, and briefly describes the AS&E BodySearch system. Operator training, x-ray image interpretation, and maximizing system effectiveness are also discussed. Finally, data collected from operation of the BodySearch system in the field is presented, and new law enforcement initiatives which have come about due to recent events are reviewed.

Keywords: BodySearch, x-ray(s), backscatter, imaging, explosives, drugs, human inspection, contraband

1. INTRODUCTION

A large global drug trade and a rising concern over terrorist activities are placing increasing demands on the law enforcement community. Over the last several years, cooperative efforts between government and industry have led to improvements and new developments of inspection technologies. The deployment of these sophisticated and effective screening devices has made it more difficult for smugglers and terrorists to get past installed security measures. However, much of this technology has been focussed on screening of parcels, luggage, vehicles, and cargo. Similar deployment of technology for human inspection has lagged behind, making the human body a more attractive vehicle for passing illegal drugs, weapons, and explosives through security check-points. For any high quality security screen to be truly comprehensive and effective, it must address all of the possible avenues by which it might be penetrated. New technological tools for human inspection are now available to help law enforcement agencies enhance their security measures. Ultimately, for these tools to be useful and effective they must not only be technologically sound, but they must also be compatible with many other requirements such as safety, convenience, and public acceptance.

The different environments where security check-points are employed have a variety of objectives. In the international customs service and border control environment, searches of parcels, cargo, and people focus on interdiction of illegal drug trafficking and smuggling of valuables to avoid import duties. In a prison environment, drug interdiction is again an objective but so to is the search and seizure of weapons or materials from which weapons can be fashioned, which includes both metals and plastics. Other security-controlled installations such as high-level government buildings, military installations, and nuclear facilities seek to prevent the unauthorized entry of weapons and explosives as well as the removal of valuable or controlled materials. One additional application for human inspection is civil aviation security. This topic has lately gained much exposure as recent world events have forced a re-evaluation of such security measures worldwide, and inspection measures considered too aggressive only a few years ago are now under deliberation. A review of some recent results of this deliberation is presented in a later section of this paper. Given the wide variety of material types, quantities, shapes, and sizes that can be considered a "threat" in any of these situations, the most beneficial tool to law enforcement officials will be one that is sensitive to detecting as wide a range of materials as possible. A tool that is sensitive to only a limited range of threats will correspondingly be of limited value.

Taking a look at some other considerations, it should be noted that in many of these applications it is the public at large that is the subject of these human inspections. Keep in mind also that the overwhelming majority of these people are honest, law-abiding citizens whose searches turn up negative. This reality requires that any human inspection method intended for routine use must meet standards of public acceptance. Issues of privacy, intrusiveness, inconvenience, and legality are therefore every bit as important as technical performance and effectivity.

There are several methods available for performing human inspections, and they vary widely in their trade-offs between effectiveness and intrusiveness. The most widely practiced methods are the least intrusive: walk-through metal detectors and pat-down searches. Because these types of searches are relatively unobtrusive, the barriers to implementation are low. However, their ability to detect a wide range of concealed materials is limited. The far more intrusive strip-search is probably the most effective search possible, but for obvious reasons it is only practicable in a restricted number of environments. Even then, profiling of potential subjects is often used as a pre-screen, with a very small percentage of the total subject pool actually being searched. Naturally, this very selective process is a limiting factor in overall interdiction effectiveness. New technological tools such as electromagnetic imaging, x-ray imaging, and trace detection are now becoming available to increase effectiveness while limiting intrusiveness. Naturally, each of these technologies has its own set of advantages, disadvantages, and trade-offs. The AS&E® BodySearch™ system, which uses backscatter x-ray imaging and is the subject of this paper, achieves high resolution images with sensitivity to both metallic (high-Z) and "organic" (low-Z) materials on the body with minimal inconvenience and intrusion to the subject.

2. THE AS&E BODYSEARCH SYSTEM

2.1 System Overview

The AS&E® BodySearch™ system, pictured in Figure 1, uses AS&E's "flying-spot" technology to scan the subject's body in two dimensions with a highly collimated x-ray beam. An extremely low x-ray exposure (5 μrem/scan) is achieved through this high collimation and through the short exposure time needed to complete a scan. The x-ray scanning platform is completely enclosed within a rugged cabinet which is capable of withstanding abusive environments and which separates the scan subjects from all moving parts of the device. The subject stands on an external stage for several seconds while the x-ray platform within the cabinet raster-scans from top to bottom. An electronic image of the subject is formed using the intensity of x-rays scattered back from the subject. This scattered x-ray intensity is a function of the atomic number and density of materials illuminated by the primary x-ray beam - which in this scenario is either the body itself or materials worn on the body. Very few x-ray interact with clothing and hair, since their density is so low, such that these items effectively disappear from the image. Denser objects such as metals, explosives, plastics, and packed drugs interact more strongly and so appear in the image along with the body itself. This can be seen in Figure 2, which shows examples of both metal and organic items on the body imaged with x-rays. The tight collimation of the x-ray beam results in high spatial resolution in the acquired images, making identification of the objects on the body easier. Note, however, that this technique only images materials on the surface of the body. It is not effective for seeing through the body or detecting materials which are concealed within body cavities. Because of this fact, two scans (front and back) are typically required for a routine inspection. Additional scans can in some cases be beneficial for identifying objects on the body. This is an area where the training and experience of the system operators can be important in maximizing system effectiveness (see Section 2.2). A more complete treatment of the system operation and features can be found in a previous publication[1].

2.2 Maximizing System Effectiveness

In the majority of cases, a simple front and back view of the person being inspected is sufficient to determine the presence or absence (and in many cases the identity) of contraband on the body. Extracting that extra percentage of detection sensitivity, though, requires slightly more. As is the case with virtually all x-ray imaging systems, the operator interpreting the acquired images is a vital component in the overall performance and effectiveness of the system. This is clearly the case with the BodySearch system. For maximum effectiveness, the operator must know not only how to interpret the images but how best to position the subjects being scanned for the best view of potential contraband. This is not a difficult task to learn, but is simply a matter of operator training. First comes an understanding of how different materials appear in an x-ray image when placed against or alongside of the body. As a rough rule of thumb, materials with high atomic number (high-Z) such as metals are good absorbers of x-rays, and so appear in the image as dark areas corresponding to few scattered x-rays. Conversely, low-Z materials (drugs, explosives, and the body itself) are good sources of scattered x-rays, and so appear as bright areas in the image corresponding to a large number of scattered x-rays. Since objects show up best when presented with a contrasting background, the most revealing view of dark, high-Z objects occurs when they are positioned in front of the bright signal from the body. Figure 3 illustrates the contrast of metals against the body. Likewise, bright, low-Z objects are easiest to detect when imaged alongside the body so that they are contrasted against the dark air background. These are ideal conditions which produce the greatest detection sensitivity. However, the operator can not always expect these conditions to occur. This is where operator training becomes so important. When low-Z objects are imaged against the body, that

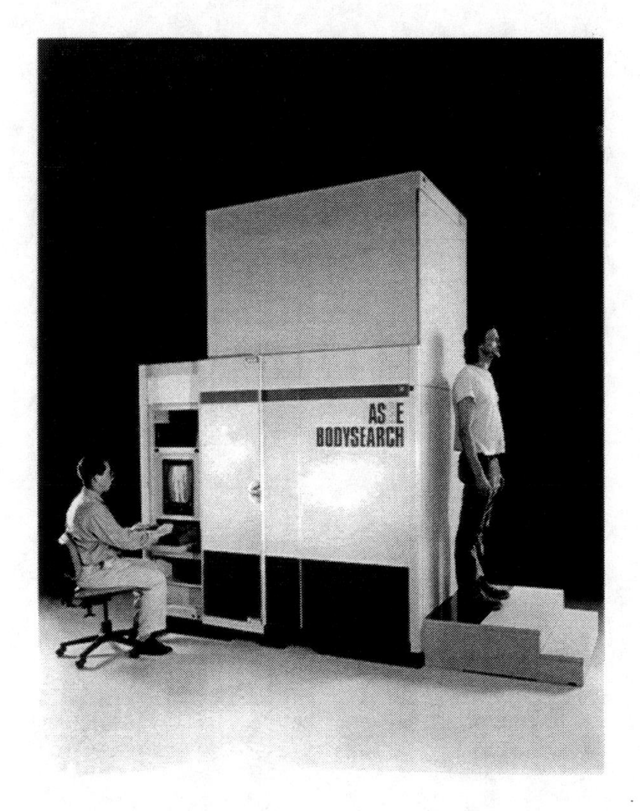

Figure 1. The AS&E BodySearch X-ray Inspection System

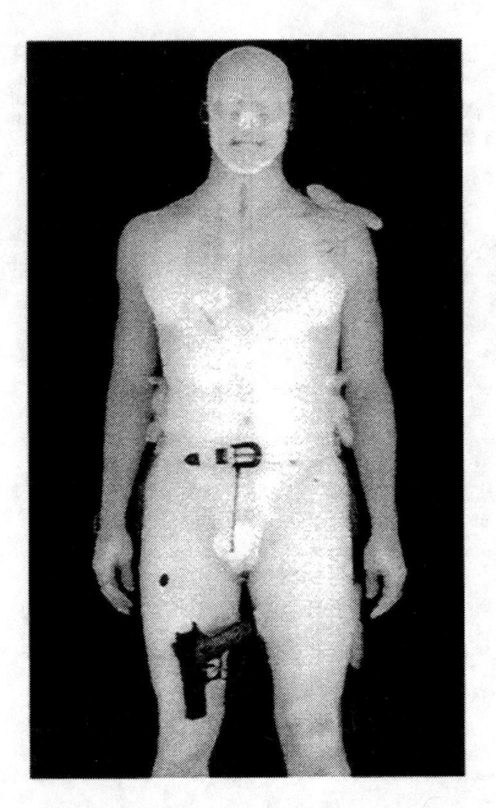

Figure 2. X-ray image taken with the BodySearch system. Visible in the image is metallic and organic contraband concealed on the body.

is, white-on-white, there are still fairly obvious indications in the image if one knows what to look for. Changes in brightness level often reveal areas that are distinguishable as foreign materials on the body. Many low-Z threats appear this way because they have a density that is different from that of the body, resulting in a difference in grey-scale in the image. Even more useful is a "shadowing" effect which usually outlines a foreign object. Because of the very small spatial extent of the primary x-ray beam, x-rays which scatter from the body immediately next to a foreign object are somewhat shielded by that object, resulting in a darker grey-scale in the image for that given pixel. This occurs all around the object and tends to outline its shape with a "shadow." Figure 4 shows an organic threat placed against the body and illustrates the shadowing effect around the object. Operators using the BodySearch system quickly become accustomed to what a normal body without contraband looks like in x-rays, such that these visual cues are easy to recognize. This is facilitated by the fact that many foreign objects on the body such as packets of drugs, molded explosives, and plastic items usually have straight edges, right angles, or uniform curvatures to their shape. These shapes are not natural on the human body and immediately draw the attention of operators who know what to look for. Once operators are trained to look for these signs, even small, subtle objects in the image can be detected which might otherwise have gone unnoticed. A skilled operator will then re-position the person being inspected so that potential low-Z threats will be presented to the system in silhouette against the air background, permitting more certain detection and identification.

This same procedure can be exercised for high-Z materials imaged alongside the body against a poorly-contrasting air background. While air produces little x-ray scatter, it produces enough to contrast against metal objects, given the right tools with which to view them. The BodySearch system is equipped with image enhancement features which, among other things, allow the operator to enhance regions of the image containing very low signal levels. This helps to take advantage of the small but useable contrast between metals and the air background making it easier to distinguish the presence of metal items. A skilled operator will use this tool and in these cases re-position the subject for imaging of the object using the body as a backdrop.

2.3 Health and Safety

As stated earlier, the BodySearch system exposes the subject being scanned to 5 μrem (0.05 μSv) per scan, which is an

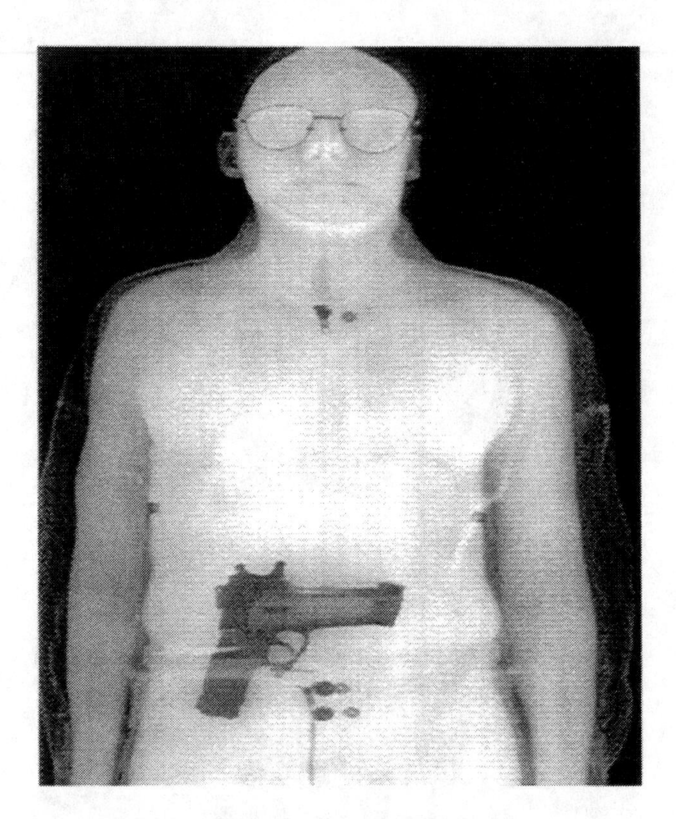

Figure 3. Metallic contraband imaged in front of the body appears with high contrast

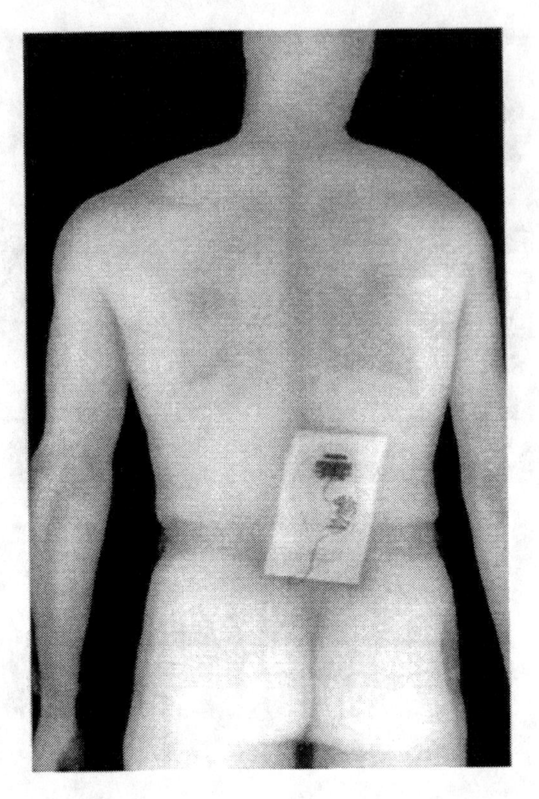

Figure 4. Organic contraband imaged in front of the body has poor contrast, but is still easily detectable due to shadowing around the object's edges.

extremely low exposure. By way of comparison, the typical natural background radiation exposure (at sea level) that we are all exposed to is approximately 550 μrem per day, or the equivalent of 110 BodySearch scans per day. Another interesting comparison can be made to the radiation exposure received during air travel. The exposure received during a round-trip flight from New York to Los Angeles is roughly equivalent to 1000 BodySearch scans. The safety of the BodySearch system for general use even in applications where multiple scans are performed on persons as an everyday routine has been confirmed by independent health physicists, and the BodySearch system has met all of the compliance requirements set forth by the U.S. FDA for entry into commerce.

2.4 Detection Capabilities

The detection capabilities of the BodySearch system are naturally dependent on the skill of the operator interpreting the images (see Sec. 2.2) and can vary depending on the body type of the subject. Because of this variation, it is not as easy to quantify absolute detection limits for the BodySearch system as it is for other types of x-ray inspection systems. To address this, AS&E has performed a series of tests to make the best determination possible of realistic detection capabilities under given conditions. These conditions are: a) the operator viewing the images has a reasonably good understanding of how different materials appear in the x-ray image, is reasonably skilled at detecting the visual cues in the images indicating the presence of contraband, and generally knows how to pose the subjects for effective imaging, b) the operator viewing the images has sufficient time (~10-20 sec) to study the images, c) the operator makes use of the standard set of image enhancement tools available on the system, d) the subject has a build representative of the norm in the U.S. population, and e) contraband items are worn on the surface of the body, not within body cavities or enclosed by tissue. Under these conditions, the BodySearch system has been found capable of detecting metallic (high-Z) threats down to a resolution of 28 AWG, making small blades, bullets, pins, and hypodermic needles detectable. Further, for organic (low-Z) threats of reasonable density (>0.7 g/cc), the BodySearch system is capable of detecting from sub-10gm quantities under ideal conditions to approximately 60gm quantities under the toughest conditions.

3. OPERATIONAL DATA FROM THE FIELD

Possibly the most valuable source of information on how the BodySearch™ system works under real-world conditions and how it will be accepted by the public at large comes from systems already operating in the field. Among the BodySearch systems currently deployed, one installation at a major Mexican border airport has proven to be such a source of data. This particular system has been in operation for approximately 1.5 years and is being used by the Drug Enforcement Division of Mexico's PGR with the objective of finding drug shipments of significant quantity concealed on persons' bodies. As passengers arrive at the airport, they are profiled by law enforcement officials. Persons considered to be likely candidates for smuggling are inspected either with a standard pat-search or through x-ray imaging with the BodySearch system. Through this process, the BodySearch system to date has effected the detection and confiscation of a significant quantity of high-valued drugs being smuggled into Mexico (exact statistics not available). Typical seizures involve 0.5 to 2.0 kilogram quantities of heroine or cocaine. Note that lower-valued drugs such as marijuana are not often found on the body either through BodySearch inspection or pat-searches, since profitable trafficking of these drugs involves quantities too large to be carried on the body. Heroine and cocaine concealments are most often found taped to the midriff or between the legs. The BodySearch system is also detecting concealments inside of shoes, with law enforcement officials specifically instructing subjects to raise their feet for imaging. This is an excellent example of how skilled operators can maximize the effectiveness of the system through proper posing of the subject.

A surprising statistic in this field example is that most of the seizures involve women, who are paid couriers. While most of the system operators are also women, same-sex matching of the operator/subject is not specifically practiced. The BodySearch system at this site is integrated into a wall, which facilitates operations. Officials do not advertise the fact that x-ray imaging is taking place, but neither do they withhold this information when asked. Despite this, and despite the fact that subjects do not know the gender of the operator viewing the images, objections from passengers are rare. One can conclude from this that use of the BodySearch system at this site has received tacit acceptance from the public. The same can also be said of a BodySearch installation in the United Kindom, where very few objections were raised by passengers during a six-month airport beta site operation. In the case of the UK site, the system was in plain view of the subjects, its method of operation was explained to every subject, and consent forms were signed by each as a routine procedure.

The BodySearch installation in Mexico also gets used in some ways that were never envisioned when the system was designed. For example, during periods of very heavy traffic through the airport, two people are sometimes scanned at once, standing side-by-side. Naturally there is some image cut-off in these cases, yet the objective of the search is still fulfilled. Also, the system is used to search briefcases and handbags by simply having the passengers hold them while being scanned. On occasion, checked luggage that is deemed suspicious is pulled from the baggage claim carousel and placed in front of the system for inspection. Although these types of inspections are not the mission of the BodySearch system, such information from use of the system in the field is extremely valuable when designing improvements and upgrades to the system performance.

4. RECENT EVENTS AND THE NRC REPORT

At the time of writing of this paper, the cause of the TWA 800 tragedy is still undetermined. Nonetheless, this event has brought to the forefront the need for improved security in civil aviation. Current human inspection techniques in use in U.S. airports have changed little since walk-through metal detectors were deployed to combat the hijacking threat of the 1960's and early 1970's. Today's inspections still focus largely on finding metallic weapons that can be used to threaten passengers and crew. This is now rapidly changing with the realization that U.S. flights are attractive targets for terrorist attacks and that a flight can be brought down with devices undetectable by simple metal detectors. Earlier this year, and before the TWA 800 incident, the National Research Council (NRC) published the findings and recommendations of a study requested by the FAA regarding airline passenger security screening. A significant part of this report focused on new technologies such as BodySearch™, and specifically addressed non-technical aspects of implementation, appropriateness, and acceptance. This report represents a new turn of events for these technologies. Previous to this year, the potential applications for which BodySearch and other related devices were considered suitable were limited to drug interdiction and restricted-population security (prisons, nuclear facilities, etc.). Now for the first time, serious discussion is taking place over using these technologies for routine security within the general population for purposes other than drug interdiction. Due to the significance of this new direction, some time is devoted here to reviewing findings from the NRC's report.

Primary issues considered in the NRC report are health, legal, operational, privacy, and convenience issues. Among these, privacy and legality are identified as some of the most important. The legal issue raised is one that revolves around the right to conduct a search. In an airport security environment, the right to search persons is restricted to a search for items which may pose a threat to the traveling public. There is no right granted to search for illegal but non-threatening items. This poses a problem, since many of the new technologies available will permit detection and identification of both threatening and non-threatening items. The report goes on to suggest that these technologies be modified to detect only threat items. Some very modest steps towards achieving this technologically are possible today, such as restricting the search zone to areas where threat concealments might be expected. This approach, though, may miss detecting some threats and will still reveal some non-threats. Truly achieving this objective is questionable and is perhaps better addressed as an operational issue rather than a technical function. The privacy issue is an obvious one. Again, an imaging system that has a high detection sensitivity will by necessity reveal more information about an individual than he or she might be comfortable with. Using computer algorithms to mask out private areas of the body (or other similar approaches) helps alleviate this problem, but at the expense of detection. Same-sex operators, availability of alternative search techniques, and operational discipline to prevent the abuse of image data are the best options available today to address the issue of privacy. Ideally, development of automated detection or computer assistance algorithms will in the future defuse these issues through system functionality rather than through operational procedures.

After reviewing the health aspects of using new inspection technologies, the NRC report concludes that these devices are generally safe for use. Because of possible problems with public perception, however, the report suggests that a public education campaign would be helpful in improving public acceptance. This is an important recommendation. In the area of public acceptance, possibly the most insightful and enlightening comment made in the report is that the amount of intrusion and inconvenience that the public will tolerate is directly dependent on the level of threat (real or perceived) and the effectiveness of the efforts to deter the threat. Certainly the perceived level of threat faced by airline travelers today is increased over past years. In the face of this threat, and coupled with the effectiveness of new technologies, the NRC conclusion indicates that deployment of devices like the BodySearch system might be greeted by a receptive public.

5. EXPLOSIVES DETECTION VERSUS DRUG DETECTION

Using the BodySearch™ system for detection of explosive threats in an airport environment is in several ways a much easier task than drug interdiction. Drugs, explosives, and the human body all have similar average atomic numbers. Many explosives, however, have a significantly higher density than the body which is seldom the case with drugs. Therefore, distinguishing explosives on the body using x-ray imaging is an easier task than distinguishing drugs due to the stronger x-ray signal they generate. Further, the minimum mass of material that constitutes a "threat" is larger for explosives than for drugs. Drug interdiction policies often target quantities as low as a gram or less (in many countries there is no lower limit for drug interdiction), while threat quantities of explosives are significantly larger than one gram. Not only is it easier to image the larger quantities of explosives, but the areas on the body where these larger masses can be concealed is more restricted. For example, the most common concealment place on the body for very small quantities of drugs is around the groin. It can be very difficult in these cases to distinguish drug concealments from the structure of the body itself. However, larger quantities of explosives are not as easily hidden, and can be expected to be worn around the midriff or on the legs. In this type of scenario, some of the computer algorithms suggested in the NRC report to enhance privacy, such as masking out the privates in an image, become more practicable for explosives than for drugs. The one image enhancement issue which is more difficult when screening for explosives is the clutter issue. When screening for drugs, a subject is usually asked to empty his/her pockets before imaging. This reduces the amount of extraneous clutter in the image that can be distracting to operators. However, in an airport security environment travelers would be scanned with pockets bulging with papers, wallets, keys, etc. This presents a different challenge to operators, who must interpret the images despite the distracting clutter.

6. SUMMARY

It is clear that in present day society there are environments in which effective screening for concealments on the human body is needed. Conventional pat- and strip-searches, while effective, are not suitable for addressing the security and drug interdiction needs that we now face due to their intrusiveness and inconvenience. Application of new techniques to this problem can help to alleviate many of the difficulties associated with human inspections. High-resolution x-ray imaging with the BodySearch™ system has been found to serve this function well. Increasing public protection from terrorist threats and enforcing the illicit drug laws that so greatly affect our society is made possible through fast, safe, and effective x-ray screening. Unfortunately we live in a world where the need for security screening of persons is likely to continue rising. Through continued development and refinement of new tools for screening, improved security need not carry such a high cost in inconvenience and invasion of privacy for the public at large.

7.REFERENCES

1. G.J. Smith, "BodySearch Technology Uses X-ray Imaging to Remove Hazards and Humiliation from Personnel Searches," Proceedings IEEE 29th Annual Int'l Carnahan Conference on Security Technology, October 1995.

The SECURE personnel screening system: field trials and new developments

Steven W. Smith

Nicolet Imaging Systems
8221 Arjons Drive, Suite F
San Diego, California 92126

ABSTRACT

Many different techniques have been investigated for detecting weapons, explosives, and contraband concealed under persons' clothing. Most of these are based on imaging the concealed object by using some sort of penetrating radiation, such as microwaves, ultrasound or electromagnetic fields. In spite of this effort by dozens of research groups, the only technique that has resulted in a commercially viable product is back-scatter x-ray imaging, as embodied in the SECURE 1000™ personnel screening system. The SECURE technology uses radiation levels that are insignificant compared to natural background values, being viewed as "trivial" and "completely insignificant" under established radiation safety standards. In the five years since the SECURE 1000 was developed, more than a dozen field trials and initial placements have been completed. This paper describes both the capabilities and limitations of the technology in these real-world applications.

Keywords: secure, security, concealed, personnel, weapons, explosives, back-scatter, x-ray, imaging

1. BRIEF DESCRIPTION OF THE SECURE TECHNOLOGY

The SECURE (Subambient Exposure Computerized Use of Reflected Energy) technology uses back-scattered x-rays to create an electronic image of the subject and any objects concealed under their clothing. A typical image created by this method is shown in Fig. 1. The system operates by directing a narrow "pencil" beam of x-rays toward the subject's body. A portion of these x-rays penetrate 1-2 cm into the body, interact by Compton scattering, and exit the body through the same surface that they entered. These back-scattered x-rays are then detected by very sensitive x-ray detectors positioned near the x-ray source. Since the fraction of x-rays back-scattered from an object depends on the object's atomic number, the detector signal is a measure of the atomic composition of the material at the location where the pencil beam intersects the body. An electronic image is then formed by moving the pencil beam in a raster scan pattern, recording the detector signal at each location. Body tissue, organic matter and plastic appear light in the image, indicating a low atomic number material. Metals, as well as bones near the surface of the skin, appear dark, indicating a high atomic number composition. Before display on the computer monitor, each image is digitally processed using various nonlinear spatial filters to provide easier image interpretation. A more detailed description of the SECURE technology is presented elsewhere[1,2].

Physically, the SECURE 1000 is 48" wide, 32" deep, 80" high, and weighs about 650 pounds. Installation of the system requires only a few minutes, with no special facilities or power required. Training of new operators typically requires about one hour of formal instruction, and several hours of hands-on experience. Persons being scanned stand about six inches in front the system, and must remain relatively motionless for three seconds. Almost immediately, the computer enhanced images appears on a display monitor. Multiple views, such as front, rear and side, require individuals to turn their bodies for additional scans.

The radiation dose imparted to each individual being scanned is three microRem. This is equivalent to the radiation dose all persons receive each five minutes from normal background radiation. Passengers flying aboard commercial aircraft are exposed to this same level of radiation about every 20 seconds. Radiation doses from medical examinations are typically in the range of 30,000 to 300,000 microRem. The SECURE 1000 has been tested by Sandia National Laboratories[3], and cleared for sale in the United States by the Food and Drug Administration (FDA).

2. RESULTS OF FIELD TRIALS AND INITIAL INSTALLATIONS

During the period from 1992 through 1996, 20 SECURE 1000 systems have been produced, three major field trials have been conducted, and 11 systems have been sold or leased to end users. Seven of these sites are within the United States. All of these test sites and installations are correctional facilities, were the primarily application is screening visitors for weapons, drugs, escape paraphernalia, and other contraband.

A major field trial was conducted by the California Department of Corrections, at the R.J. Donovan Correctional facility, between April 1994 and November 1995 (20 months). R.J. Donovan is a medium security California prison near the Mexican boarder, housing approximately 5000 male inmates. The SECURE 1000 was used to screen 100% of all visitors, averaging 200 visitors/day, 5 days/week. The screened individuals were 70% female, 15% male, and 15% children. During the course of the 20 month trial, over 100,000 persons were screened.

Detailed statistics were compiled for the first ten weeks of the trial. During this initial period, 9,621 visitors were screened, resulting in 109 (~1%) of the persons being subjected to a more invasive hand search. This typically involved a same-sex security officer escorting the person to a private room to identify the object seen in the SECURE 1000 image. This secondary search seldom required more than a few minutes. Prohibited items were found in 17 cases, including: over $2,000 in currency, prescription drugs, safety & hat pins, pagers and cosmetics (which could be used to hide illicit drugs).

In addition to these items directly detected, the system was clearly a deterrence to smuggling. During the first ten weeks, five balloons containing illicit drugs were found in the visitor waiting area, presumably discarded out of fear of being detected. These included: heroin, methamphetamine and marijuana. In addition, ten (~0.1%) of the visitors refused to be scanned, and were required to leave the facility. Given the other statistics in this study, it is reasonable to assume that some or all of these persons were attempting to smuggle illicit drugs, and were deterred by the SECURE 1000.

From the final report of the R.J. Donovan staff, the SECURE 1000 was found to be "a valuable tool in deterring visitor smuggling ... excellent in detecting weapons, escape paraphernalia, money, and jewelry ...

an excellent morale booster ... (and) motivated staff to do the best job possible." Concerning the limitations, the staff concluded that the system "falls short in regard to narcotics detection." This comment was particularly directed at the gram or subgram quantities of drugs typically smuggled in a prison setting, and the inability of the technology to detect objects hidden in body cavities. Based on the results of this field study, the California Department of Corrections Technology Transfer Committee approved the SECURE 1000.

3. RADIATION SAFETY

As little as five years ago, the overwhelming view of both the security and radiation protection communities was that ionizing radiation would never be used for searching persons. This long-held belief was challenged by the development of the SECURE technology, and its subsequent clearance by the FDA. The issue of x-ray use can be broken into two categories, the *actual* safety of the system, and the *perceived* safety of the system.

The *actual* safety of the system is easy to address. Every person receives 250 to 500 microRem of radiation dose each day from naturally occurring radioactive materials in the air and soil. Variations in this level result from local conditions, elevation, housing materials, and even clothing. For example, wearing a sports coat instead of a sweater produces a few percent change in the background radiation reaching the body. Over a period of a few hours, this results in a few microRem change in dose. In this regard, being screened with the SECURE 1000 is literally equivalent in risk to wearing a sports coat instead of a sweater, or eating lunch on an outside patio instead of an inside dining room. The SECURE 1000™ is safe, because its risks are no larger than the risks from daily activities and behaviors that are considered *unconditionally* safe. These are well accepted principles of radiation protection, and there is no controversy in this area.

Critics of the SECURE technology have maintained that the general public would strenuously object to the use of x-rays, regardless of the dose. They argue that the *fear of radiation* will cause an unacceptable number of people to refuse to be screened, or worse, initiate lawsuits. If anything has been learned in the last four years, it is that this concern is completely unfounded. To date, there have been no lawsuits, or even the threat of legal action. Subject compliance is excellent at these sites, with only a very small fraction of individuals refusing to be screened. It should be emphasized that the subjects in these trials are visitors, not incarcerated persons. This provides a strong indication that other security applications involving the general public would have similar results. Of course, the possibility of legal action always exists, and a certain fraction of the public might refuse to be scanned. However, the fear of radiation is clearly not a significant operational problem.

4 PRIVACY ISSUES

The factor which limits the widespread use of the SECURE technology is objection to the images displayed. This is very gender specific; nearly all males being screened accept the display, while nearly all females express significant concern over the invasion of privacy.

This issue was successfully overcome in the test sites by relocating the display monitor into an adjacent room. Two advantages are provided by this configuration. First, the subjects being screened are kept unaware of how the system operates, and therefore cannot devise strategies on how to defeat the technology. Second, the personal interaction between the subject and the security officer viewing the image is severed. The previously cited statistics on subject compliance show the effectiveness of this solution in the correctional setting. Another proposed solution is to require that only female security officers screen female subjects. This was not tested in the field sites because of the effectiveness of the remote monitor; however, it may be useful in other types of security applications.

4. CONCLUSIONS

The SECURE technology was introduced to the security community five years ago. Since that time, field trials and initial installations have shown it to be a valuable tool for screening visitors entering correctional facilities. The public's reaction to the use of x-rays can be described as *uneventful* and *quiet*. While privacy concerns are an issue with the technology, at least for visitors entering correctional facilities, they are manageable with appropriate operational procedures.

The SECURE 1000 has been accepted as an effective tool for contraband detection in the correctional arena. Screening members of the general public who visit prisons and jails has set the stage for use of this technology in other high security applications. This includes passenger screening at airports, employee screening at nuclear facilities, and employee and general public screening at government buildings,

5. REFERENCES

1. S. W. Smith. "Detection of objects concealed under person's clothing using the SECURE system," *Proceedings of the First International Symposium on Explosive Detection Technology*, pp. 261-268, S.M. Khan, editor, FAA Technical Center, Atlantic City International Airport, Feb. 1992.

2. United States Patent Number 5,181,234

3. B.T. Kenna and D.W. Murray. "Evaluation tests of the SECURE 1000 scanning system," SAND 91-2488 UC-830, April 1992. Prepared by Sandia National Laboratories for the United States Department of Energy.

Figure 1. A typical front image acquired with the SECURE technology. Several concealed objects can be detected, such as: 1/2 pound of simulated C4 plastic explosive, handgun, screwdriver, two coins in the pocket, snaps on the trousers, metal watchband.

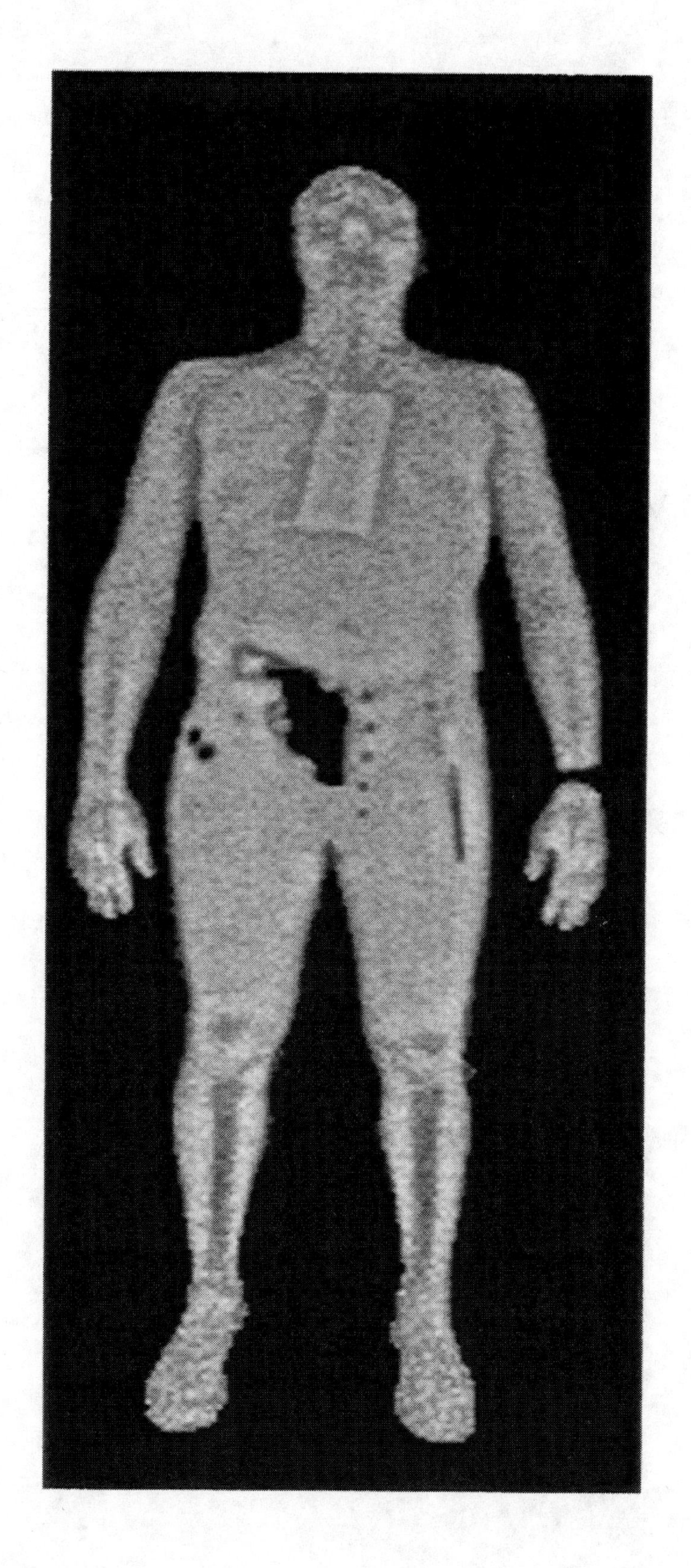

Part B

POSITIVE IDENTIFICATION: TESTING METHODS AND TECHNOLOGIES

SESSION 4

Positive Identification: Trends and Technologies

Trends in AFIS Technology: Past, Present, and Future

Guy Cardwell
Behnam Bavarian

Printrak International
1250 North Tustin Ave., Anaheim, CA 92807

ABSTRACT

Automated Fingerprint Identification has a history of more than 20 years. In the last 5 Years, there has been an explosion of technologies that have dramatically changed the face of AFIS. Few other engineering and science fields offer such a widespread use of technology as does computerized fingerprint recognition. Optics, computer vision, computer graphics, artificial intelligence, artificial neural networks, parallel processing, distributed client server applications, fault tolerant computing, scaleable architectures, local and wide area networking, mass storage, databases, are a few of the fields that have made quantum leaps in recent years. All of these improvements have a dramatic effect on the size, speed, and accuracy of automated fingerprint identification systems. This paper offers a historical overview of these trends and discuss the state of the art. It culminates with an overview and educated forecast on future systems, especially those "REAL TIME" systems for use in area of law enforcement and civil/commercial applications.

Keywords: Fingerprint, AFIS, IAFIS, Real-Time, Computer Architecture

1. History Of AFIS Innovations

The following information provides the historical background on the 20 years of technology development that form the foundation for the latest, standard-setting sixth-generation AFIS products:

First-Generation (1975)

Installation of the first fingerprint reader at the FBI in 1975. Research into pattern recognition and minutiae matching was begun in 1969. It is interesting to note that the first introductory publications in pattern recognition start appearing around 1974[1,2] from premier researchers in academics, while in the fingerprint image processing industry very sophisticated and accurate algorithms were being integrated into products.

System 250S and System 300 (1978 and 1982)

The second and third generation systems laid the foundation for today's Real-Time AFIS with capabilities such as automatic pattern classification and space(rotation and translation) invariant matching algorithms during searching.

System 400 (1985)

The fourth-generation system introduced several advance features:

1. First use of gray-scale images in display in user friendly Graphical User Interface allowing image manipulation and image restoration techniques.

SPIE Vol. 2932 • 0-8194-2334-3/97/$10.00

2. First to offer side-by-side image display during search verification for expert viewing and case preparation for evidence
3. First jukebox implementation, image storage and retrieval systems using optical platters and Library Handlers for automatic optical platter selection and mounting.

ORION (1987)
The fifth-generation AFIS used DEC minicomputers and workstations, VAX and MicroVAX computers and Digital-specific operating systems (VMS and RSX-11). Advances characteristic of ORION included:

1. Introduction of advance image processing and feature extraction algorithms the pioneering Contextual Enhancement Processor (CEP) And its hardware realization, a 28-board massively parallel distributed computer architecture with over 3 Billion Operation per second computational power.
2. First to deliver DCT based gray-scale image compression and decompression in hardware.
3. Ability to interface third-party livescan devices to AFIS using additional hardware and software.
4. Ability to interface on-line with computerized criminal history systems.
5. Improvements to matcher accuracy and speed (MXD and Turbo Matcher modules).
6. Ability to use Rapid Changers (linked "cartridge-style" jukeboxes with larger optical disk capacity) rather than larger, more unwieldy, less reliable library handler solutions that stored less per optical disk.

Series 2000 (1994)
The sixth-generation products are "pushing the envelope" for AFIS once again. Advances characteristic of Series 2000 include:

1. Open systems software following POSIX standards
2. Complete ergonomic design in physical layout and use of the latest man-machine interface paradigms and standards for design of the GUIs and the peripherals; Workstation screen design and furniture developed in cooperation with human factors experts to ensure user comfort and productivity in the highly intensive tasks characteristic of AFIS operations.
3. Scaleable architecture built upon COTS products, supporting easy networking, TCP/IP, and interface to other systems.
4. Invention of the Intelligent Image Processing algorithms.
5. Fingerprint Processor 2000, which replaced 28 boards with a single board at the size of a floppy disc drive with SCSI interface. Some of the characteristics are:
   - World's Fastest Image Processing Board (2 Billion Operations per Second)
   - Full Programmability Allows Algorithm Enhancements; Field Upgradeable
   - A Computational Engine for High Speed Implementation of
     - Intelligent Image Processing Software on a Single Board
     - Minutiae Matching Algorithm
     - WSQ Compression/Decompression
     - Image Quality Verification
     - Mugshot Image Processing and Compression
   - Reduces Environmental Requirements Designed around the Texas Instrument TMS320C80 Multi-Media Video Processing (MVP) Chip; The Flagship member of the industry-leading DSP (Digital Signal Processing) chip.
   - MVP Sets New DSP Performance Threshold:

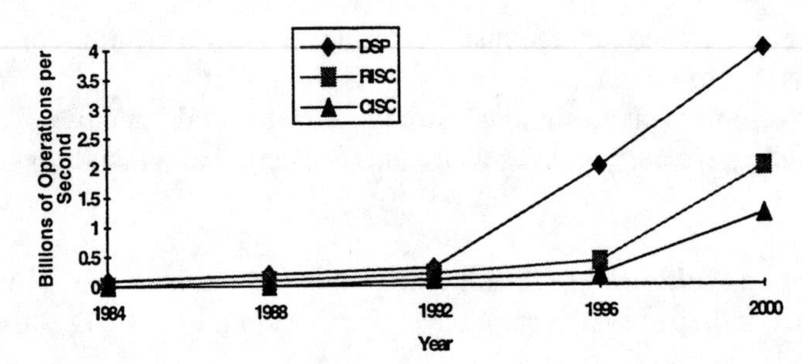

Figure 1. The comparative performance of the DSP, RISC, and CISC architectures.

6. Compliance with National and International Standards:
   - ANSI/NIST -CSL 1-1993 Data Format for the Interchange of Fingerprint Information
   - FBI/NIST compliant WSQ Compression and decompression
   - FBI compliant Live scan technology
7. Minutiae Matcher 2000; DRAM storage of minutiae data and parallel search processing, allowing the MM 2000 to operate at full capacity for all types of searches which improved search speed drastically setting the stage for Real Time Integrated AFIS solution.
8. Scaleable, parallel, distributed, and fault tolerant search processor with computational power equivalent to supercomputers, supporting Real Time positive identification solutions in both law enforcement and commercial markets.
9. Fault-tolerant RAID magnetic storage, which dramatically improved image retrieval times and increased system reliability
10. Fully Integrated LiveScan technology with AFIS Functionality. The LiveScan Stations are designed "from the ground up" as AFIS point of entry devices, not just card creation units. They offer a higher level of integration flexibility and more AFIS pre-processing capability than can be obtained anywhere in the industry. Because AFIS pre-processing is available at the LiveScan Station. The new generation of these units offer fingerprint-based data integrity checking, AFIS quality evaluation and rescan prompts, and FBI-certified WSQ compression — all crucial components of fully integrated, real-time AFIS-livescan operations.

2. Real-Time Integrated Automated Fingerprint Identification System;

The Future Trend

The future trend of AFIS technology goes hand in hand with the innovations and break through from the computer technology and advances in the computer vision and artificial intelligence algorithms. The themes of AFIS is already set around

- Real Time solution, and
- Integrated Total Solution.

The following example defines what such a system look like. It is an AFIS installation in the State of Louisiana[3], the first Real-Time Integrated Automated Fingerprint Identification System. The high level functions of this system are:

1. To provide a statewide integrated AFIS using "State of the Art" technology for seamless connectivity and search response times in seconds.

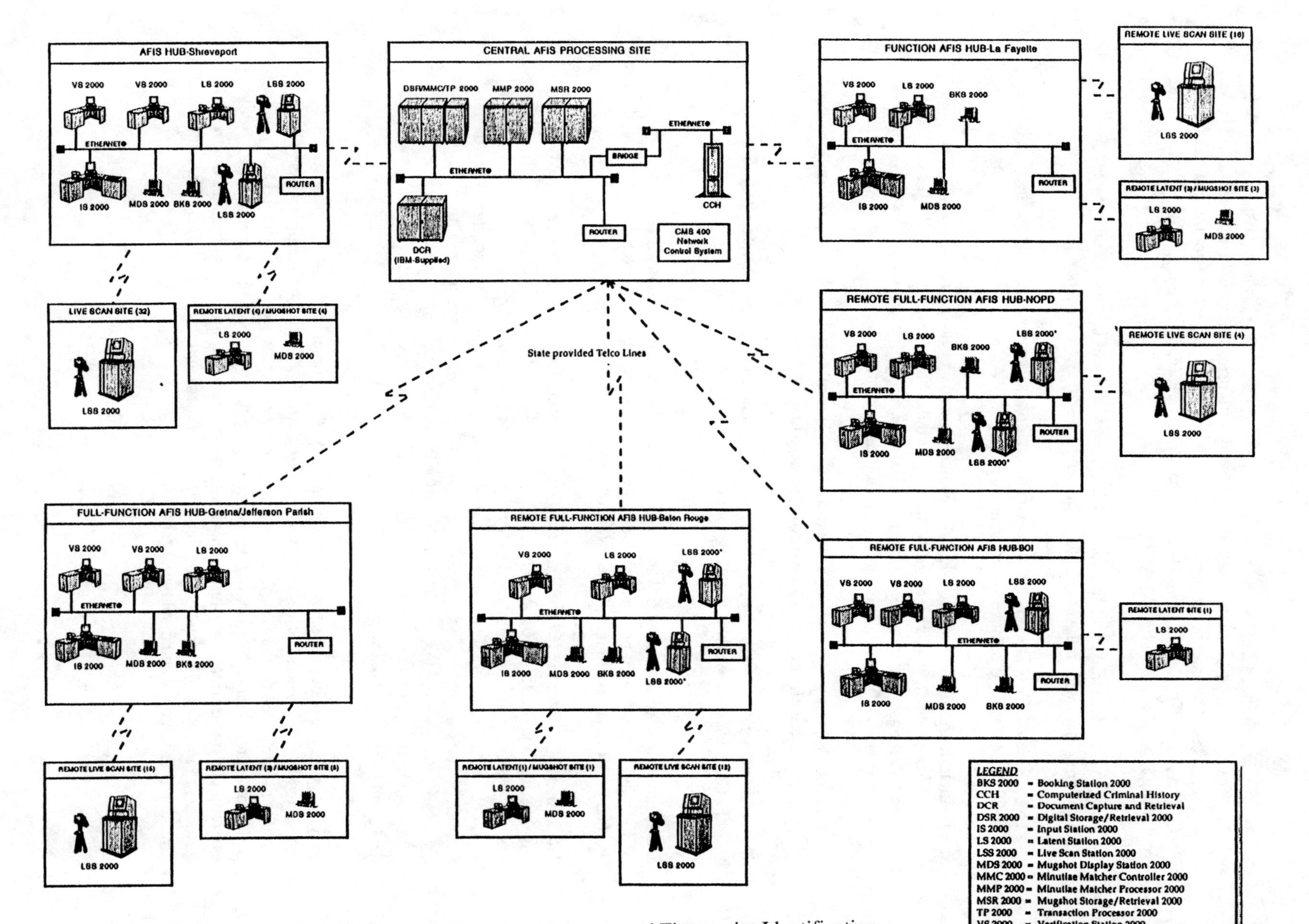

Figure 2. Louisiana State Police Real-Time Integrated Automated Fingerprint Identification System

2. To provide on-line access to the State's CCH (Computerized Criminal History) database for timely and accurate identification of persons with prior criminal history records. This is accomplished by matching tenprint fingerprints with criminal records in the State Police's CCH and NOPD(New Orleans Police Department) MOTION (Metropolitan Orleans Total Information On-line Network) systems.

3. To provide additional identification capabilities to the local law enforcement agencies by providing them with mugshot photographs and line-up creation capabilities.

4. To provide a DCR (Document Storage and Retrieval) System that will interface with the AFIS to capture and retrieve fingerprint card images.

5. To include installations at the DPS&C (Department of Public Safety & Corrections, Data Processing Center, with six (6) Full-function Remote hubs, seventy-eight (78) remote live scan and mugshot capture locations (a total of 84 Live Scan Stations (LSS 2000)), and approximately 20 remote latent and mugshot display sites.

6. To provide managed Local Area Networks (LAN) at all installation sites, and multiprotocol routers linking all sites via a state-provided network.

7. To support network management by an existing Code Management System (CMS) 400/CMS 6000 network monitoring facility.

The Central Site AFIS is located at the DPS&C Data Processing Center. This site will make AFIS, mugshot, and document image databases accessible to Full-Function Remote sites. It also provide network access to the CCH server for all workstations connected to the AFIS.

The system allows for six (6) Full-Function Remote locations. AFIS workstations, Mugshot Display Stations (MDS 2000), and LSS 2000s is installed at these Full-Function Remote hubs. The stations support fingerprint card entry, fingerprint live scan entry, quality control, search verification, and mugshot display. Each Full-Function Remote provides services for a defined number of booking sites. These services include: receiving input from booking sites, pre-processing all searches and forwarding to the Central Site, receiving and reviewing search results, returning search results to the booking sites, and accepting updates of AFIS and CCH data from the booking sites.

Seventy-eight (78) booking sites and approximately 20 latent and mugshot display sites are connected by Wide Area Network (WAN) circuits to the Full-Function Remote sites (These circuits are provided by the State). Each of these sites is connected to an Ethernet LAN and a LAN-connected router. The LAN enables existing booking systems to exchange data. Booking sites capture fingerprints, mugshots, and demographic data and interact with Full-Function Remotes to initiate and complete tenprint searches. Latent and mugshot display sites enter and initiate latent searches and display mugshots.

3. Concluding Remarks

The last 20 years of successful field proven AFIS technology is the testimony to the fact that use of fingerprints is the only viable solution for positive identification as compared to recently developed interesting biometrics such as hand geometry, iris scan, retinal scan, or face recognition which is in its infancy.

As the result of a new wave of interests in the positive identification market and new large law enforcement AFIS procurements, a very large number of small and big companies became experts in AFIS technology over night, with chain of claims about new algorithms and technologies, mostly re-inventing the wheels or based on small closed laboratory experiments. Unfortunately, this is adding to the hype and it is confusing serious developments and transfer of technology to new sectors. On the top of this, one has to also consider the social and political implications associated with use of fingerprints. However it is paramount for an industry interested in using fingerprints for added level of security and audit to evaluate the traditional law enforcement AFIS technology and plan a smooth transfer of technology to meet its application requirements. This include observation and analysis of the operation of installed AFIS sites in a law enforcement agency for the few AFIS vendors which have been delivering the technology for many years.

AFIS is a very complex and challenging field embracing many areas of mathematical techniques and solutions and advance computer technologies. The science of fingerprints is over one hundred years old and the AFIS technology is only twenty years old. There is a lot of room and need for innovations and improvements in the technology which has to be done on the existing solid foundation to make it a successful industry.

4. References

1. Tou, J. T. and Gonzalea, R.C., *Pattern Recognition Principles*, Addison-Weley, Reading. Mass. 1974.

2. Fu, K. S., *Syntactic Methods in Pattern Recognition*, Academic Press, New York, 1974.

3. "*State of Louisiana Request for Proposal, AFIS Procurement*", Document number B20460ED, January 1995.

NC-TEST; Non-Contact Thermal Emissions Screening Technique for Drug and Alcohol Detection

Francine J. Prokoski

MIKOS Ltd.
Fairfax Station, Virginia 22039-7025

ABSTRACT

Drug abuse is highly correlated with criminal behavior. The typical drug-using criminal commits hundreds of crimes per year. The crime rate cannot be significantly reduced without a reduction in the percentage of the population abusing drugs and alcohol. Accurate and timely estimation of that percentage is important for policy decisions concerning crime control, public health measures, allocation of intervention resources for prevention and treatment, projections of criminal justice needs, and the evaluation of policy effectiveness. Such estimation is particularly difficult because self reporting is unreliable; and physical testing has to date required blood or urine analysis which is expensive and invasive, with the result that too few people are tested.

MIKOS Ltd. has developed a non-contact, passive technique with the potential for automatic, realtime screening for drug and alcohol use. The system utilizes thermal radiation which is spontaneously and continuously emitted by the human body. Facial thermal patterns and changes in patterns are correlated with standardized effects of specific drugs and alcohol. A portable system incorporating the collection and analysis technique can be used episodically to collect data for estimating drug and alcohol use by general unknown populations such as crowds at airports, or it can be used for repetitive routine screening of specific known groups such as airline pilots, military personnel, school children, or persons on probation or parole.

KEYWORDS: drug screen; drug usage estimate; non-attribution testing; drug effects detection

1.0 DRUG AND ALCOHOL ABUSE IN THE UNITED STATES

Estimates of the number of drug users in the U.S. vary widely. The Office of National Drug Control Policy (ONDCP) estimated in its 1990 report that there were 860,000 hard-core cocaine addicts in the US. At the same time, the Senate Judiciary Committee announced their estimate of 2.2 million hard-core cocaine addicts. Billions of dollars in government funding at the federal, state, and local level are being allocated each year on the basis of such weakly-supported and politically-tuned estimates. Even more tragically, opportunities to improve lives through rehabilitation and treatment may be lost because inconclusive statistical evidence is considered for the effectiveness of such measures. Estimate errors could be reduced by using significantly larger sample sizes, if techniques were available for low cost massive sampling.

While developing better and cheaper sampling techniques, the primary need is to increase the effectiveness of programs for the reduction of drug use, under existing operational constraints. Constraints include legal issues (invasion of privacy), resources (money, trained people, labs, time), and technology (time to process drug tests and chain of custody considerations.) Drugs are changing and usage patterns are changing. There is a need for flexible techniques which can spot trends which may be associated with new drugs and changing patterns of drug use in the population at large, prior to the development of specific tests which meet evidentiary requirements.

Urinalysis has taken more than 20 years to achieve acceptance in the criminal justice system. Hair sample analysis still has not received general acceptance after an equal time. If we consider also fingerprint and DNA analysis, we must conclude that it takes multiple decades to develop forensic technologies acceptable for

SPIE Vol. 2932 • 0-8194-2334-3/97/$10.00

evidentiary use. Meanwhile, however, we can assist law enforcement agencies with early warning techniques, which supply probable cause for allocation of limited resources to specific cases, and which may later be accepted for evidentiary use.

NC-TEST provides statistical, realtime estimates of drug and alcohol (D&A) usage by large populations of unknown persons. It offers an unprecedented chance to assess and track D&A use among populations in specific areas without personal privacy concerns. In particular, High Intensity Drug Trafficking Areas (HIDTA) can be continually monitored to determine the short term effects of any changes in drug policy, supply level, distribution patterns, or types of drugs available.

The automated non-attribution system provides statistical estimates of the percent of persons who pass the system's cameras who are chronic or recent users of any vasoactive drug, including specific classes of drugs such as cocaine, heroin, marijuana, or alcohol. The systems makes the determination on classification of each person, based upon the degree of similarity between specific features of that person's thermal signature and reference thermal signatures obtained under known D&A use conditions.

The system can be used to determine differences in the percentage of persons entering and subsequently leaving schools, the workplace, and public facilities who appear to be under the influence of drugs or alcohol. When a significant differential exists, it may serve as an indicator of substance use within the premises, and may constitute probable cause for searches, or scanning with attribution, to be conducted within those facilities.

2.0 CURRENT ESTIMATION TECHNIQUES

Current statistical estimates, while useful, are not sufficiently accurate nor timely to support policy makers. Sometimes conflicting estimates result from various indicators and surveys, due to their limitations and ranges of measurement.

National Indicators which are routinely compiled include:

- The Drug Abuse Warning Network (DAWN) reports emergency room admissions and drug-related deaths. The data is not lab confirmed, and may be underreported unless drug relevance is clearly paramount. DAWN covers only non-Federal hospitals.
- Uniform Crime Reports are filed by only the most serious crime in cases of multiple charges. One report per incident is filed, not one per person.
- The System to Retrieve Drug Evidence (STRIDE) seeks only to assess price and purity, with no distinction made between wholesale and retail and no reference to the persons involved.
- The Drug Use (DUF) program performs urinalysis on arrestees whose cooperation is voluntary, and matches results against their self-report. It places priority on sampling arrestees charged with non-drug-related offenses, and therefore underestimates the drug using proportion of the arrestee population.
- Client-Oriented Data Acquisition Process (CODAP) considers only those admitted to federally funded drug treatment programs, and is limited by the number of slots available.

Major Drug Indicator Surveys include:

- The National Household Survey, Epidemiological Catchment, and High School Senior Survey rely on self-reporting and so can at best assist in assessing trends. They are affected by changes in the stigma of admitting drug use. About 25% of potential participants refuse to cooperate, which is a measure of the potential underreporting error. These surveys do not include estimates associated with high risk groups which are not members of households, or who either dropped out of high school prior to the senior year or were absent on the day of the survey.

- The National Ambulatory Medical Care Survey, and National Hospital Discharge Survey rely on physician and hospital reporting on drug and alcohol abuse diagnosis and treatment. This data is biased by the exclusion of high risk groups who do not routinely seek physician or hospital assistance.

Due to the small sampling sizes of current surveys, and the types of limitations mentioned, various organizations extrapolate the same data and come to widely varying conclusions. For example: the National Clearinghouse for Drug and Alcohol Abuse in Rockville MD estimates that roughly 10 million workers are current abusers and estimates cost to US industry in lost productivity is $7.2 billion. However, the National Institute for Drug Abuse in Bethesda, MD has suggested a $60 billion per year cost in accidents, tardiness, absence, drug-related theft, and medical costs. Other experts suggest even higher cost figures. Reliance upon self reporting has been found to have high error rates, since the individuals routinely lie about the frequency and extent of their drug usage. Urinalysis has been found to be accurate enough, but too intrusive and too expensive for mass routine use.

Evidence has been collected on the progression from occasional to frequent use, and from one drug to another. For example, many heroin users report progression from initial use to daily use in less than 1 month. Crack cocaine users have a similar escalation rate; powder cocaine users have much slower escalation. The dynamics of drug use within segments of the population can change faster than current sampling techniques can track.

The proposed thermal screening technique offers the potential of more accurate and more timely estimates of drug use in the general population, as well as in subgroups, due to its ability to collect massive amounts of data without intrusion or contact with the sampled population, at very low cost, frequently or continuously.

3.0 BACKGROUND

Psychoactive drugs are substances which alter mood, thought processes, or behavior, or which are used to manage neuropsychological illness. Drugs may be classified according to their mechanism of action, but often the mechanism is not known. Drugs can be classified by chemical structures, but similar structures produce different pharmacological activity and dissimilar drugs induce the same activity. Therefore, the most useful classification is by the most characteristic behavioral effect or widest clinical application. Using that criterion for classification, some of the classes of drugs of strongest interest are:

- Depressants, such as inhalants and alcohol
- Psychomotor stimulants such as cocaine and caffeine
- Narcotics such as morphine, codeine, and heroin
- Psychedelics and hallucinogens such as LSD, marijuana, hashish, and cannabis.

The classification may be dependent upon dosage: alcohol is a general central nervous system (CNS) depressant at high dosage, but at low doses may cause behavioral excitation. Some drugs affect several areas of the brain, such as those involving temperature regulation, behavioral arousal, satiety, or rage

There are four stages of drug action: administration and absorption of the drug into the body; distribution of the drug within the body, interaction of the drug with its receptors in the body, and elimination of the drug from the body. Drug absorption refers to mechanisms whereby the drug passes from the point of entry into the bloodstream. The timing depends on: route, dose, and dosage form. Administration most commonly is in one of five ways: orally, rectally, by injection, by inhalation and by absorption.

In an average size adult, each minute the heart pumps 5 liters of blood, which is roughly equal to the total amount of blood in the circulatory system. Therefore, once a drug is absorbed into the bloodstream, it is rapidly distributed throughout the circulatory system, generally within the one minute period. Drug responsiveness is a function of the size and weight of a person, as well as his metabolism and many other factors such as drug interactions, level of physical activity, etc.

Drugs move predictably throughout the body and drug movement increases in speed once the drug circulates in the bloodstream. Blood picks up drugs from receptors and moves towards the heart. It then goes to the lungs to obtain oxygen and release carbon dioxide. Then it goes back to the heart before returning to tissues via arteries, carrying the drug along with it.

Although drug effects may vary among people, the majority of persons experience the same type of reaction, and have the same anatomical involvement for a given drug, including general symptoms: high temperature, increased blood pressure, asymmetrical temperature, sweating or not, cold extremities, pupil dilation, constricted veins and arteries.

Among these "universal" effects for specific drugs are:

- Inhalants dilate blood vessels thereby causing reduced blood pressure, flushed feeling and dizziness.
- Alcohol is a CNS depressant. It dilates the blood vessels in the skin, producing a warm flush and a decrease in body temperature.
- Psychostimulants such as cocaine and amphetamines constrict blood vessels, produce excitement, alertness, euphoria, reduced fatigue and increased motor activity. Blood pressure and heart rate increase, pupils dilate, blood flow shifts from skin and internal organs to muscle, and oxygen levels rise as does glucose level in the blood.
- Amphetamines increase blood pressure, stimulate the CNS, and cause bronchial dilation. Methamphetamine (ICE) is smokable, speeding up the onset of the effects.
- Psychostimulants include caffeine and nicotine. Caffeine has a slight stimulating effect on the heart, increasing cardiac contractility and output and dilating the coronary arteries. However, it exerts an opposite effect on the cerebral blood vessels: it constricts these vessels, thus decreasing blood flow to the brain.
- Nicotine from cigarettes, oral snuff, chewing tobacco and nicotine gum produces elevated blood nicotine levels for 120 minutes after use and causes increased psychomotor activity in the cerebral cortex. It is a stimulant, with a period of depression following. Normal doses cause increased heart rate, blood pressure and cardiac contractility. It initiates vasocilation, increasing blood flow to meet the increased oxygen demand of the heart. Blood pressure and heart rate increase.
- Marijuana is an unique sedative-euphoriant-psychedelic drug; The active ingredient (THC) is a sedative and analgesic, which decreases spontaneous motor activity, decreases body temperature, increases pulse rate but has little effect on blood pressure. Dilated blood vessels produce bloodshot eyes, changes in blood pressure, corresponding dilation and constriction of veins and arteries, and related changes in level of activity of drug receptor sites. All are accompanied by minute temporary changes in local temperatures.

The temperature of skin directly over a superficial blood vessel is always warmer than skin which is not so located. Thermal cameras can detect these minor heat variations and translate them into different levels of grey in the output signal. When there are vasoactive changes at a specific site due to drug activity, then there are changes to the detected thermal signatures. The level of sensitivity of the infrared camera will determine which of those changes can be detected. Metabolic activity associated with eating, exercise and other normal functions can in general be distinguished from changes induced by drugs. Drug-induced changes have different cycles of build up and decay, and affect specific areas of the brain and therefore specific pathways of the vascular system.

Current IR cameras, operating in the mid to long wave region of 3-12 microns wavelength, can image facial patterns caused by superficial blood vessels which lay up to 4 cm below the skin surface. Current infrared cameras are sufficiently sensitive to be used as screening tools to detect possible drug and alcohol usage within scanned populations. Future developments are expected to further refine the detectability and specificity of the D&A screening procedures. Future cameras will have increased sensitivity which should translate into imaging of deeper thermal signatures and more sensitive models of the body's thermodynamic behavior in response to drug use.

4.0 SCREENING TO DETECT DRUG AND ALCOHOL USE BY UNKNOWN INDIVIDUALS

Thermal (infrared) imagery collected passively from the face can provide detection and classification of thermal signatures which are highly correlated with the use of drugs and alcohol. The substances of greatest interest, namely illicit drugs, generally produce significant thermal changes in the facial thermograms, with those changes decaying away and the normal facial thermogram re-emerging over time as the effect of the drug diminishes. The rate of decay is a function of the recency and amount and type of substance used. In chronic abusers, permanent abnormalities in the facial thermogram may develop which persist even weeks after drug use. Similar results pertain to alcohol abuse.

Any substance which changes the metabolism of the body affects the heat distribution of the body and to some extent the facial thermogram. Figure 1 shows the results of alcohol on a facial thermogram. Substances such as cocaine, marijuana, and heroin also cause significant changes in the skin temperatures at specific locations. Different substances cause different sets of changes. These changes decay with time, depending on the amount, type, and purity of substance used and how often the person uses it. Mixtures of substances produce combined results.

Any ingestion, inhalation, or administration of a vasoactive substance in sufficient quantity to produce significant changes in cognitive or motor activity will also produce thermal changes in blood pressure and blood flow within certain portions of the vascular system. Evidence of such effects will be manifest within the facial thermogram, given a sufficiently sensitive thermal imager is used. Using the NC-TEST technique, the existence of abnormal thermal signatures can be detected at specific precise facial locations which are interrelated to specific areas of the brain.

Given that a suitably sensitive and well resolved infrared camera is used, and that a suitable image processing and change detection technique is employed, a subject person may be declared to be free of significant quantities of any vasoactive substance on the basis of showing no static or dynamic abnormalities in his thermal signatures which match those found to be associated with drug or alcohol effects.

Persons showing no thermal abnormalities, or who best match substance-free thermal signatures in the reference database, may be classified as substance-free. Persons showing significant thermal abnormalities may be classified as substance-involved, and be further classified into the group whose thermal signatures are the closest match when compared against reference thermal signatures obtained from clinical trials in which known substances are administered to subjects whose resulting thermal signatures are then recorded, analyzed, and/or compared to statistical results obtained from persons or populations classified by prior drug or alcohol use. The accuracy of the NC-TEST procedure can be compared against accepted blood and urine testing standards.

When NC-TEST is used in a non-cooperative non-attribution mode for population analysis, the system's method includes processing of each frame of imagery to find each face in the frame, test for sufficient facial area to permit classification of the face, testing to select only facial images of sufficient quality, tagging the face to track it through subsequent frames, assuring that the same tag is applied to each appearance of the same face in succeeding frames, obtaining time varying thermal signatures from each tagged face, and processing a batch of different persons' sequences of images in order to produce statistically significant estimates of D&A usage from the population sampled. The face tracking processing allows for maximum analysis time of each face which appears in the field of view of the thermal camera, and yet assures that each person is counted only once, regardless of the time spent within the field of view. Both steps serve to improve the accuracy of the population screening estimates.

5.0 POSSIBLE ERRORS

Certain prescription or over-the-counter drugs may cause thermal effects which are similar to those caused by certain illicit drugs. Therefore, the thermal analysis technique is currently considered only a screening test, rather than definitive proof for the presence of specific drugs.

The most relied upon method of drug testing is Urinalysis, which is considered highly invasive. Urinalysis testing results in 2.0% to 7.5% false positives, which can be reduced through the use of a medical review officer (MRO), who analyzes all positive test findings, makes judgments on them against the test subject's medical profile, and renders a verdict. Reasons for MROs' changing the test conclusion include consideration of the use of prescription drugs or non-controlled substances that register as controlled substances on both first-screen immunoassays and the more sophisticated and expensive gas-chromatography tests. First-screen tests identify chemical combinations which may or may not be controlled or illegal substances. Gas-chromatography testing is more specific and more comprehensive, but also more expensive.

It is expected that the false positive rate for facial thermal screening will be no worse than that produced by urinalysis. Data on use of prescription and over the counter medications may be factored into estimates obtained from the NC-TEST system, to possibly reduce the false positive error. Such data is generally considered less prone to error due to lack of stigma regarding self-reported use of legal substances, and availability of corroborating information from medical sources and from business documents.

6.0 LEGAL AND PRIVACY ISSUES

The NC-TEST system provides real time screening and classification of every person passing in front of the system camera, without requiring that they cooperate or even be aware of the system operation. Since the proposed screening need not include identification of the individuals, or recording of their thermal or visual images, legal challenges based on Fourth, Fifth, or Fourteenth Amendment rights should not be obstacles.

Thermography is color blind; skin tone cannot be directly determined from facial thermograms. This is considered a positive feature when concerns are raised about differential treatment for various ethnic groups.

Technical, legal and policy issues are involved in each of the current techniques used for drug and contraband detection, and have impact on new initiatives for curbing drug presence in schools and workplaces. Since NC-TEST provides general population screening without attribution, it should generate less resistance to its use. The current legal guidelines for screening with attribution provide an indication of the possible restrictions which may be levied on use of the NC-TEST system in schools, which is the application of greatest interest.

6.1 Searches.

Officials in public schools are bound by Fourth Amendment restrictions in conducting searches on school property. The US Supreme Court in New Jersey v. T.L.O., 469 US 325, 105 S. Ct. 733 (1985) held that the warrant and probably cause requirements which apply to police-initiated searches do not apply to searches conducted by school personnel. However, a school search must be: justified at its inception, based on reasonable grounds for suspecting the search will reveal evidence of a violation of law or school rules, and reasonably related in scope to the circumstances which justified interference in the first place. Id. at 342.

Although probably cause is not required in searches by school personnel, the degree of reasonable suspicion necessary to conduct a search does vary with the degree of intrusiveness of the search. The standard adopted in T.L.O. depends on the circumstances surrounding the search. Therefore, the standard of Fourth Amendment reasonableness needed to search a locker or a pocketbook may fall short of the reasonableness for a strip search.

Sniff searches using dogs to conduct general searches of the outside of lockers and cars has been upheld on the theory that a person does not have a reasonable expectation of privacy in the air surrounding a manmade object in a public place. A sniff is not a search. A harder question is whether the sniff of a person by a drug-detection dog is a search. In a series of cases in the 1980's, the courts held that a sniff of students by drug-detection dogs is a search and is unreasonable unless based on individualized suspicion supported by specific and articulable facts. Horton v. Goose Creek, 690 F.2d 470 (5th Cir. 1982). That same theory, however, does not apply if the sniffer is a person. Teachers have been allowed to sniff students' hands to detect marijuana in Burnham v. West, 681 F. Supp. 1160 (E.D. Va 1987).

6.2 Metal detectors.

In People v. Dukes, 151 Misc.2d 295, 580 N.Y.S.2d 850 (Crim. Ct. 1992) the use of metal detectors at Washington Irving High School in Manhattan was upheld. Among the actions taken by school systems who prevailed in the court cases, prior notification to the students that searches would be conducted, requiring three activations of the metal detector prior to search, and consistent use of the scanning technique without arbitrary selection of students were all considered important.

6.3 Urinalysis.

At least one court has upheld drug testing of student athletes under the Fourth Amendment. Schaill v. Tippercanoe County School Corp., 864 F.2d 1309 (7th Cir. 1988). The court noted that students voluntarily submitted to the drug testing and that refusal to submit resulted only in the loss of a minor privilege, namely the right to participate in athletics.

In Acton v. Vernonia School District 47J, 796 F. Supp. 1354 (D. Or. 1992), a district court emphasized that the school district presented hard evidence that student athletes in the district were using drugs, noting that "amorphous statistics or generalized notions about the national drug problem would not have justified the searches". The court also noted that the urinalysis program was the least intrusive means of decreasing drug use by student athletes. However, on appeal, the Ninth Circuit Court of Appeals reversed the district court, ruling that prevention of injures and deterrence of drug use by high school athletes were not sufficiency important governmental interest to justify the invasion of privacy involved in random drug searches by urinalysis. The U.S. Supreme Court agreed to review the Acton decision, and subsequently upheld the legality of the drug testing program.

6.4 Surveillance.

Use of cameras and two-way mirrors in public areas of a school campus, including school buses, hallways, and classrooms has been upheld by a variety of courts, when installed in public areas where there is no reasonable expectation of privacy. When installed in locker or restrooms, differing legal opinions have resulted.

6.5 Tips.

Students are entitled to due process before being suspended from school even if they engaged in violent or illegal conduct. Often, searches by school personnel are based upon tips from other students. The accused student has the presumed right to cross examine the tipster in court. The school system cannot assure the safety of the tipsters, but

Courts have held that the cloak of anonymity is essential to obtaining student assistance in notifying school authorities when they witness criminal activity. Newsome v. Batavia Local School District 842 F.2d 920 (6th Cir. 1988). Reliance on tips from students opens a school system to highly disruptive tensions and legal battles..

6.6 Comparison to NC-TEST analysis

Thermal imaging of an individual is far less intrusive than strip searches or urinalysis, and even less intrusive than metal detector scanning to detect contraband weapons. The thermal detector is merely collecting and focusing heat generated spontaneously by the human, whereas the metal detector is itself radiating energy towards the human and then detecting the interference caused by dense metallic objects such as guns. The energy levels used for metal detector scanning of humans is very low, but still presents some concerns with respect to safety and health. Thermal imaging, by contrast, is totally passive, and the thermal cameras can be considered merely a subset of standard video cameras, where a different frequency of light (namely heat energy) is being imaged. Metal detector searches have been upheld by the Supreme Court as being minimally intrusive, where the reasonable expectation of privacy is not subject to the discretion of the official in the field. Similarly therefore we can argue that the use of thermal imaging cameras to detect possible drug-related thermal emissions is functionally analogous to the use of metal detectors to detect weapon-related reflected emissions.

7.0 NC-TEST ANALYSIS METHOD

7.1 Generic facial thermograms

The vascular system supplying the human face typically exhibits thermal variations on the order of 7°C across the facial skin surface. Certain general features, such as hot patches in the sinus areas, relatively cool cheeks, and cold hair pertain to all facial thermograms. Other features such as specific thermal shapes in certain areas of the face are characteristic of a particular person. Measured disturbances to other features, such as the general symmetry between two sides of a face, range of thermal variations in the forehead, peak temperature, size of the canthi pattern, and variations in those disturbances over time, may be correlated with a high probability of drug or alcohol use. No previous study has considered the application of thermal imaging as a screening tool; although clinical trials on headache pain and cerebrovascular disease have noted the effect of classes of drugs and alcohol on the facial thermograms.

Variations in temperature across the facial surface can be imaged by thermal cameras sensitive to wavelengths in the 3-5, 8-12, or 3-15 micron ranges. Current cameras available commercially can provide thermal resolution of better than 0.07°C and spatial resolution of better than .02", resulting in 65,000 to 265,000 discrete thermal measurements across the surface of the face. For most such cameras, that thermal map is regenerated 30 times per second to produce either a video output or a direct digital signal.

The vascular system has a common structure in each person, with known pathways for instance from the heart to the brain, and known pathways between blood vessels in the face and those in the brain as shown in Figures 2 and 3. Using the NC-TEST technique, through clinical drug trials using known types, amounts, purity, and administration techniques, the thermal effects over time at specific facial locations and the effect of varying the protocols can be observed. Since the thermal effects may be quite small and localized, it is key to utilize the same precise locations in each subject. That provides repeatability of measurements over time without requiring the application of registration markers to the face, or the use of invasive techniques to repeatably find the same locations. Also, it provides a method for comparing corresponding locations in different subjects. MIKOS has patented and patent-pending methods for repeatably locating corresponding vascular sites, which are utilized in the current developments.

Statistical analysis of the time-varying thermal signatures at each facial site before, during, and after drug or alcohol administration provides a reference dataset which represents the thermal effect of that substance under the protocol used. A library of substance effect signatures is being developed for various drugs and other substances for which screening is desired, including prescription and over the counter medications, tobacco, and alcohol. Figure 4 presents an illustration of thermal signatures associated with NC-TEST locations in the face. Also, non-substance signatures associated with substance-free subjects can be developed for the general population, or stratified by sex, age, size, medical history, or other characteristics of the substance-free subjects.

7.2 Clinical trials to produce references

After an individual ingests a drug, changes in his thermal signature gradually occur until a thermal "climax" is reached after which the signature gradually returns to its normal state. In chronic drug users, permanent physiological changes may occur such that there is no longer a smooth total decay of the apparent drug-induced effects. By processing a significant number of thermal images, thermal signature markers are identified and related to standardized vascular system locations whose thermal variations are highly correlated with use of the particular substance. References may be developed for an individual, for a class of individuals grouped by age or other characteristic, or for a general population.

A general determination of substance-free vs. substance-influenced classification may be made based upon data collected on the thermal effects of various substances of interest. In addition, certain substances produce characteristic results which may be identifiable from detailed analysis of the thermal signatures associated with facial landmarks, and/or with distribution statistics from those signatures. Furthermore, the rate of change at any point in time may be a discriminator between chronic and recent use of each drug. Techniques for processing sequences of thermal images may enhance the visibility of bilateral asymmetries, anomalous static conditions, and unusual time-varying trends in the thermal signatures.

Using currently available thermal imaging cameras, we now have the ability to use thermal signatures associated with specific landmarks locations in the face to indicate activity levels of specific arteries in the brain which are known to be affected by particular drugs. Therefore, when a substance is known to affect particular functions, vascular pathways from the face to the corresponding brain areas should be analyzed for related thermal signatures at landmarks points along the pathways, under the assumption that increased activity at the brain site will be found to correlate with increased vasomotor activity along pathways leading to that site, as evidenced by thermal changes.

7.3 Standardization of references

In order to best compare images from different people and under different conditions, facial thermograms are standardized and registered to vascular landmarks in the face. A subset of the landmarks which relate to blood vessels or areas of the face affected by a particular substance is selected. These "substance active" landmarks are thereafter considered in any tests for that substance. The use of more sensitive infrared cameras can increase the total number of landmarks detectable (including ones associated with deeper blood vessels) and thus the detectability of D&A effects.

The thermal effect of a particular substance on each of the substance-active landmarks is shown by a time-varying signature of apparent (or actual) temperature vs. time as in Figure 4 for alcohol. The time period involved can be two frames, or a longer period of multiple frames. The temperature values may be normalized or not, whichever provides better demarcation of the substance presence.

7.4 Substance use markers

The collection of differences between the time-varying thermal signatures for the substance-active landmarks, compared to the collection of time-varying thermal signatures for the same locations in the absence of the substance, represents the marker for that substance. For each substance of interest, a marker may be developed for a particular individual, for a class of persons grouped according to some criterion, or for a more general population. The substance-free marker can likewise be developed for a particular individual, class of persons, or general population.

7.5 Detection of substance use

In subsequent screening of a known individual for a particular substance, his current thermal image(s) are analyzed to extract substance-active landmarks which can be seen in the available image(s). The set of thermal signatures is compared to the substance marker collection and substance-free marker collection. Measures of similarity are calculated for individual, class, or general population comparisons for each substance of interest. Various correlation methods can be used for comparing a collection of thermal signatures to collections associated with substance use, or with substance-free references in the system library.

A measure of goodness of match is made between the collected thermal signatures and the signatures for each substance under each protocol in the library. The system manager selects a threshold to be applied to each such comparison, such that matches which are closer than that threshold will cause the system to issue a notice of a possible substance detection.

7.6 System Output

The results of comparisons with the different markers may be recorded or stored or output to decision makers. Alternately or in addition, thresholds may be automatically applied to the calculated differences to render a pass/fail or clean/under-influence determination. The statistical estimate of confidence in the determination can also be presented.

The system is deployable for programmable periods of time, during which it will analyze and classify each face which appears within its field of view. The system does not routinely record or store the thermal images, although provision are made to do that during testing and evaluation of the system in order to allow for improvements to be made in the system and compared with earlier results. The output from the system includes graphical results similar to that shown in Figure 5. The Cumulative Detection Index of the y-axis represents the number of persons who the system estimates have used marihuana, cocaine, or heroin in an amount and within a timeframe which results in a residual level indicated by the x-axis value at the time of the analysis. The x-axis represents the confidence level of drug signature indication, which is related to the detection precision of the testing and analysis procedures. Separate curves indicate the specific drugs detected, and a composite survey indicates detection of any of the substances. Due to the frequent use of combinations of drugs, the composite curve is expected to be more significant than its components.

The system is to be tested using known populations of drug users, and its results will be compared to urinalysis results. The comparison will be used to select thresholds for system decisions on classification of thermal signatures prior to deployment.

8.0 SUMMARY

MIKOS solicits the support of other medical, imaging, and law enforcement support organizations in developing and applying the NC-TEST methods to an analysis of our nation's drug problem. Society has a responsibility to its citizens to protect them from crime, while also respecting individual privacy. Non-attribution deployment of the NC-TEST system is designed to provide statistically valid data, to measure and monitor drug usage by populations at low cost, in real time without intrusion. No competing technology exists.

Society has a related responsibility to rehabilitate criminals. The recidivism rate for drug users who commit felonies is high. About half of felony drug offenders placed on probation in 1986 were rearrested for a new felony within 3 years. However, if probation is disallowed, other undesirable effects occur, such as crowded jail conditions causing harm to inmates and staff, and expensive resulting litigation. As the number of persons on parole and probation continues to increase, largely as a result of mandatory sentences for drug offenders, better techniques are needed to select participants for release programs and establish monitoring conditions for their release. While the initial proposed use of NC-TEST is for population estimates without attribution, the system can be used very effectively to detect drug and alcohol use by known individuals who have been enrolled into the system. That application is the topic of a separate future publication.

9.0 REFERENCES

1. F. J. Prokoski et al., Method and Apparatus for Identification of Individuals and their Conditions from Analysis of Elemental Shapes in Biosensor Data Represented as N-dimensional Images, U.S. patent 5163094, November 1992.

2. F. J. Prokoski, Method and Apparatus for Flash Correlation, Patent Application U.S. Serial No. 08/314,729; Notice of Allowance received.

3. F. J. Prokoski, Registration of Medical Images using SIMCOS; presentation to MEDTEC, Medical Technology Conference of the National Security Industrial Association, Orlando, Florida, 9-11 July 1996.

4. Frank H. Netter, M.D., Atlas of Human Anatomy, Ciba Pharmaceutical Division 1989; Figures 2 and 3 are adapted from color Plates 17 and 131 and are reproduced in reduced grey scale version with the approval of Ciba-Geigy Corporation, Summit, New Jersey.

5. Office of National Drug Control Policy, The National Drug Control Strategy: 1996; Executive Office of the President, Washington D.C. 20500. 6. Bureau of Justice Statistics, Drugs, Crime, and the Justice System; December 1992, NCJ-133652.

6. Gesina L. Longenecker, PhD., How Drugs Work: Drug Abuse and the Human Body; Ziff-Davis Press, Emeryvile, CA. 1994.

7. Robert M. Julien, M.D., PhD., A Primer of Drug Action, W.H. Freeman and Company, New York, Seventh Edition 1995.

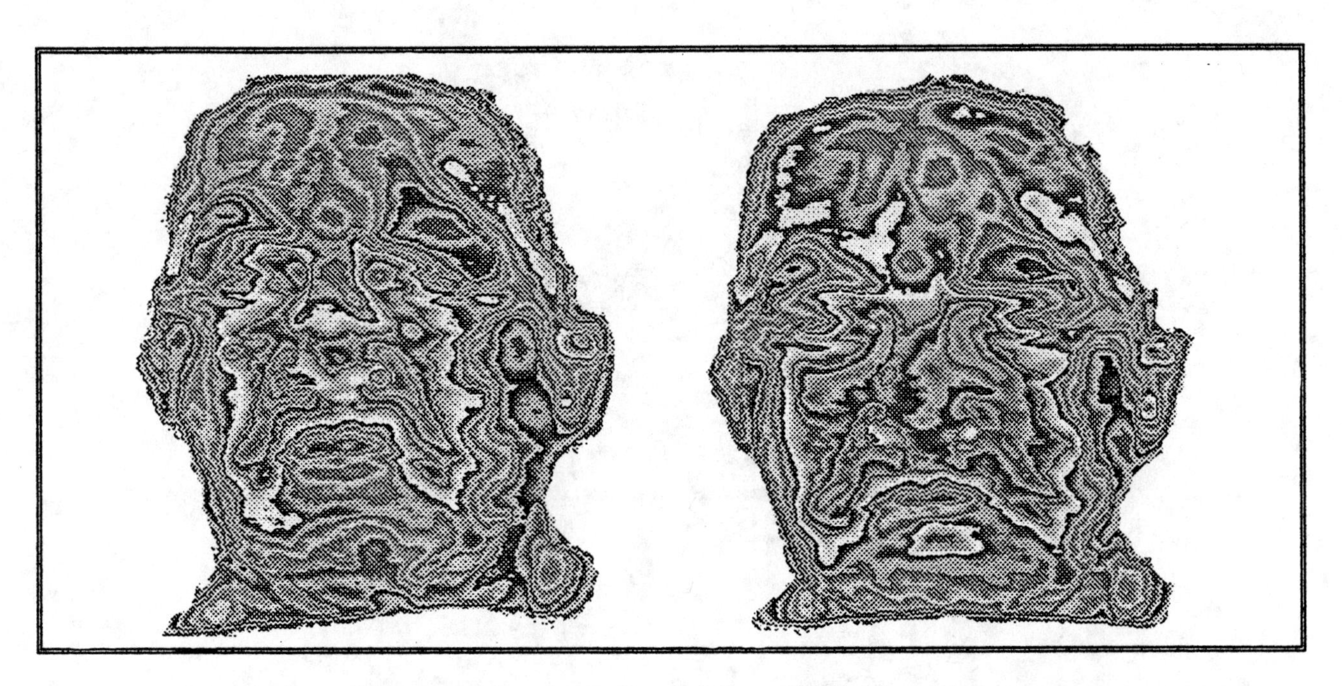

Figure 1 (a) and (b) Pre and Post-Alcohol Facial Thermograms

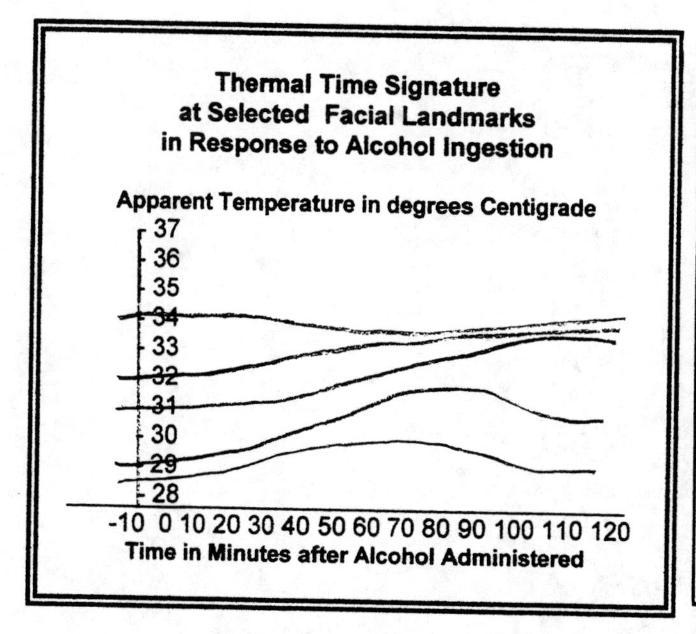

Figure 4 Thermal Signature at sample Landmarks

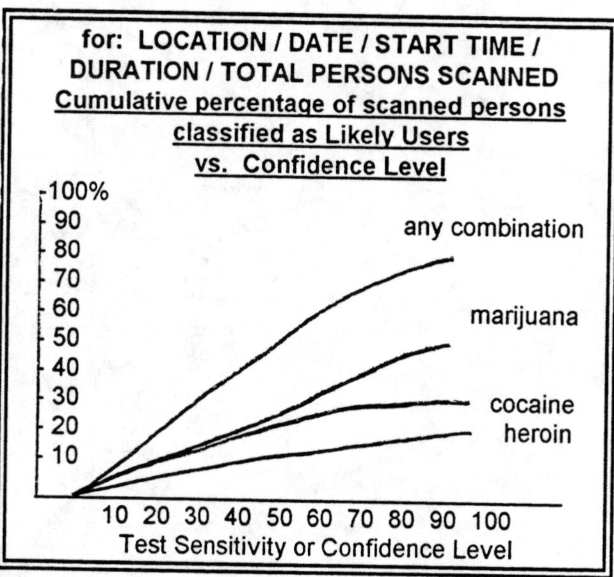

Figure 5 Sample System Output

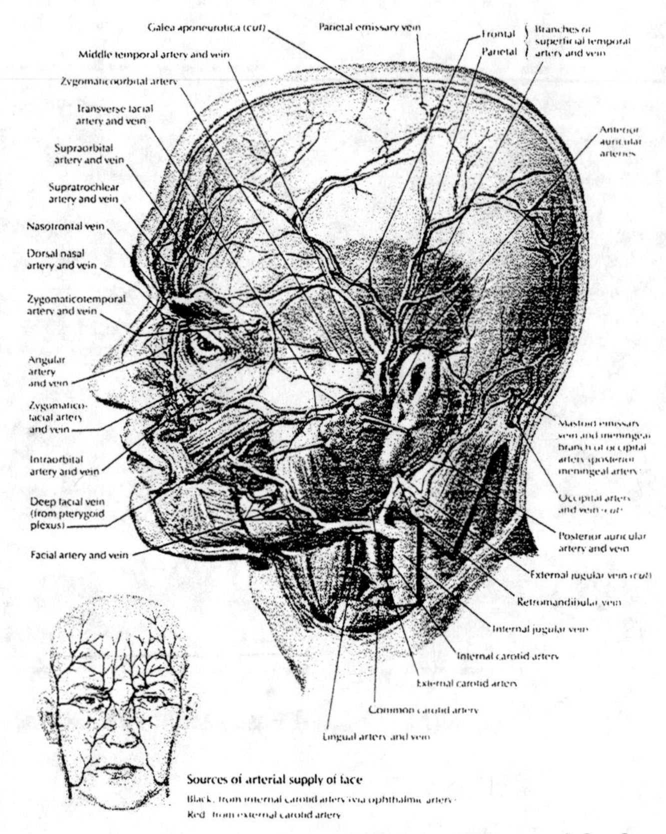

Figure 2 Superficial Arteries and Veins of Face and Scalp

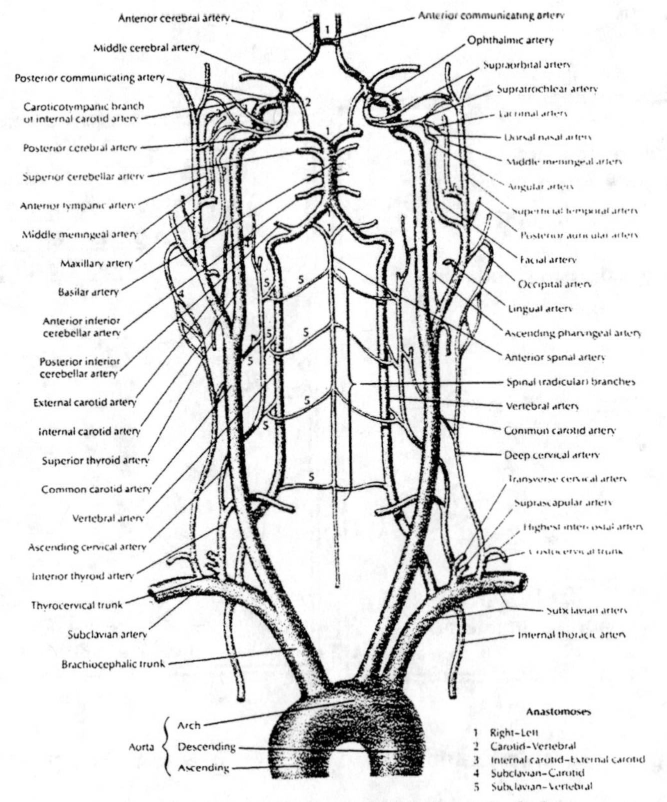

Figure 3 Arteries to the Brain: Schema

SESSION 5

Biometrics: Trends and Technologies

Directional enhancement in the frequency domain of fingerprint images

Richard Pradenas

Printrak International
1250 N. Tustin Ave., Anaheim, CA 92807

ABSTRACT

One method to enhance or analyze fingerprint images (a member of the class of oriented images which everywhere have dominant local orientations) is to use a directionally applied enhancement operator. Critical to this operator is the correct local direction. Typically, the local direction is found by correlating lines of constant extent but different angles with the local gray level intensity. The extent of these directions is a trade-off between performance in high-curvature and noisy regions. If the extent is too short the correlation works well in high-curvature regions but fails in noisy ones. The longer extent does well in noisy regions but fails in high-curvature ones. This paper shows how the directional image enhancement can also be done in the frequency domain. This approach overcomes the noise and high-curvature limitation of the spatial domain methods.

Keywords: digital image processing, frequency domain, directional image enhancement, fingerprint images

1. INTRODUCTION

Fingerprints are part of a class of oriented images characterized by dominant local directions [1]. This paper presents the development of a method for enhancement of fingerprint images in the frequency domain that uses this local direction information.

For fingerprint processing the enhancement is essential due to the non-stationary and low frequency noise caused by over-inking, under-inking, non-uniform pressure when rolling prints, and lighting variations. The enhancement processes should not change the ridge structure. It is usually followed by binarization and thinning. Minutiae, bifurcations and ridge endings, are detected on the thinned image. Simple non-directional enhancement methods such as histogram modification, bandpass filtering, homomorphic filtering, and local area contrast enhancement do not make use of the large amount of information available in the local directions (Reference [2] contains a good survey of image processing techniques as applied to fingerprints.). Using this information, although computationally expensive, pays off in better accuracy and higher hit rates, especially in large minutiae databases [3]

The Fourier Transform is part of a general theorem of transforms where a signal can be decomposed into a set of orthogonal basis functions [4] [5]. The Fourier transform uses sine and cosine waves as the set of orthogonal basis functions. (Other transforms, such as the Walsh [6] with its "square wave like" set of basis functions might be more applicable for fingerprint decomposition and enhancement, especially at the local level.) Moving from the spatial domain to the frequency domain and back is done with the forward, equation 1, and inverse, equation 2, Fourier transforms.

$$F(s) = \int_{-\infty}^{\infty} f(x) \exp(-i2\pi x s) dx \tag{1}$$

$$f(x) = \int_{-\infty}^{\infty} F(s) \exp(i2\pi x s) ds \tag{2}$$

SPIE Vol. 2932 • 0-8194-2334-3/97/$10.00

For sampled images the forward, equation 3, and inverse, equation 4, 2D Discrete Fourier Transform (DFT) are used to move between the two domains:

$$F(u,v) = \frac{1}{N}\sum_{u=0}^{N-1}\sum_{v=0}^{N-1} f(j,k)\exp\left(\frac{-2\pi i}{N}(uj+vk)\right) \quad (3)$$

$$f(j,k) = \frac{1}{N}\sum_{u=0}^{N-1}\sum_{v=0}^{N-1} F(u,v)\exp\left(\frac{2\pi i}{N}(uj+vk)\right) \quad (4)$$

The forward DFT gives the amplitude, phase, and frequency of the sinusoidal components than make up an image. The inverse DFT states that the original image can be reconstructed from these components. The reversible relationship between the forward and inverse transforms tells us that what is done in one domain will be reflected in the other.

Directional image enhancement, like many signal and image processing tasks can be done in either or both domains. The unique characteristics of each problem dictates where it can best be solved. Filtering can be done as a convolution in the spatial domain or as multiplication in the frequency domain. Background patterns are discernible in the spatial domain and are easily removed in the frequency domain. A sinusoidal signal in broadband noise might be difficult to see in the time domain but is readily visible in the frequency domain.

Directional enhancement in the spatial domain requires a good estimate of local direction and an enhancement function that incorporates this direction. Typically, local directions are found by correlating pixel intensities along a set of pre-defined directions. The direction with the lowest variation in intensities or the highest correlation is selected as the local direction. The physical extent of the correlation is a compromise between performance in noisy areas (over and under inked and scarred areas) and performance in high curvature regions (core and delta regions found in most fingerprint patterns). A long correlation does well in low curvature regions and operates well in the presence of noise. A short correlation does well in high curvature regions but performs poorly in high curvature regions. Noisy areas have low correlation in all directions and are difficult to distinguish from high curvature regions. Scarred, cut, and creased areas are also problems as these artifacts have high intensity values with little variation along their direction and look like a valid directions. This local direction can be used for image enhancement by an enhancement function that reduces the intensity variations along the local directions and increases them in the orthogonal direction. The result is an image that can be easily binarized. In the case of fingerprint images the enhancement process (in either domain) improves the gray level separation between ridges and valleys and drives the image from gray scale towards binary.

The proposed enhancement method is different than two other published frequency domain directional enhancement algorithms.

The first method [7] is described as a Fourier short-space adaptive wedge filter. Short space because the algorithm operates on the 2D DFT of 32 x 32 pixel sub-images. The spatial domain sub-image is enhanced by attenuating the unwanted frequency components. The spectral image is shifted so that the DC component is centered. The wedge filters are centered at DC and divide, in angle, the spectra. Each filter has a pre-defined angular orientation and extent and overlap each other by 50%. The wedge filter with the maximum energy is found. The unwanted spectral components in this wedge filter are reduced and the inverse DFT taken.

The second method [8] uses 16 of directional bandpass filters set at pre-defined directions to form 16 directionally filtered images. The filtering is done in the frequency domain. The enhancement is done in the spatial domain at the pixel level. Each enhanced pixel is the weighted average of pixel values from "N" pre-filtered images closest in orientation to the local ridge orientation (found in the spatial domain). N, the number of pre-filtered images that are averaged, is a function of the distance from the pixel to a core or delta. More pre-filtered images are combined when the pixel is close to these singularities to improve accuracy in these high curvature regions. Not discussed in the paper is how the core and delta locations are found (the best estimate of core and delta locations are done on enhanced images). The authors note that finding local ridge flow is difficult in noisy regions. Their averaging of images near cores and deltas also indicates that determining local ridge flow is difficult in high curvature areas.

2. FREQUENCY DOMAIN DIRECTIONAL IMAGE ENHANCEMENT

The enhancement algorithm discussed here makes use of the unique spectral characteristics of fingerprint images. Directional enhancement is done in the frequency domain by finding the frequency components that contribute to the underlying ridge structure, amplifying these components, and the moving back to the spatial domain.

There are a number of significant differences between the proposed algorithm and the adaptive wedge filter and the directional bandpass method. Local direction in the proposed algorithm is found from the local 2D DFT and is not constrained to be one of a set of pre-defined directions. Unwanted spectral components in the wedge filter are attenuated while the proposed method isolates the low frequency noise components from the components that form the ridge structure and these components are amplified. Neither core nor delta locations are required.

2.1. Frequency domain characteristics of fingerprint images.

Fingerprint images because of the strong local orientations have different frequency domain characteristics than other images. Figures 1 illustrates these differences with the log normalized magnitude of a 512 point 2D DFT of a fingerprint image and a face of a Madrill. Both spectra are shifted so that DC is near the center. Most of the spectral energy of the fingerprint is concentrated in an annulus centered at DC. The spectral energy of the Mandrill is almost uniform.

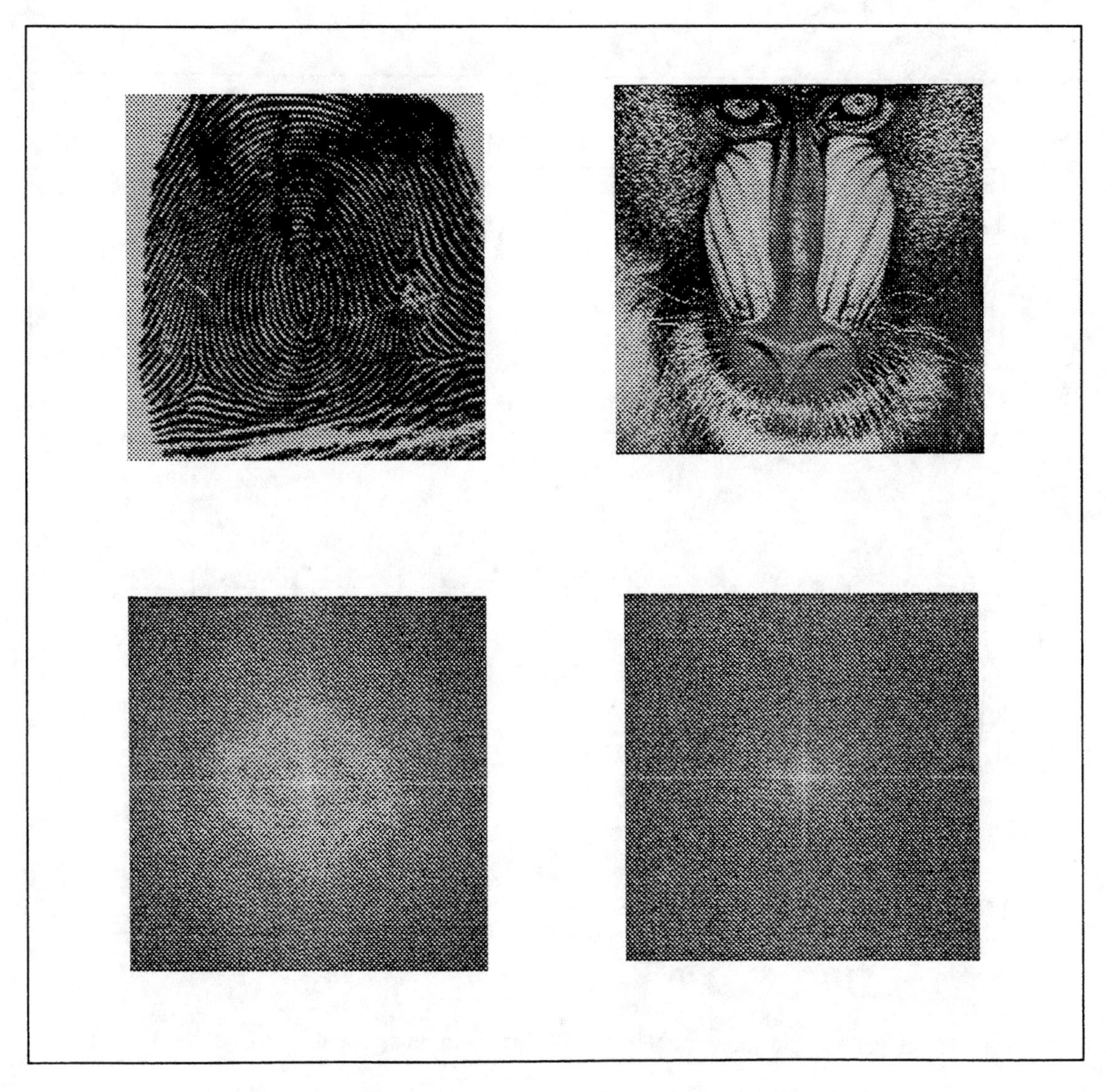

Figure 1. Spatial and Spectral Images of a Fingerprint and a Mandrill

Figure 2 is another comparison of the spectral energy for fingerprints and the Mandrill. This figure shows the average radial energy of 4000 fingerprint images in the NIST Special Database 4 [9]. A 512 point 2D DFT was used to move to the frequency domain and the spectra shifted so that DC is near the center. The vertical axis is in relative dB; both images were normalized by the maximum value. The horizontal axis is wave number which is analogous to frequency and is the radius from DC. The radial energy is the sum of energies at a given radius. Most of the energy in a fingerprint is found between wave numbers 30 and 70 and peaks at about wave number 50. These wave numbers are dependent on the size of the DFT (If a 1024 point 2D DFT were used the same peak would occur at wave number 100). The energy content, away from DC, for the Mandrill is fairly constant.

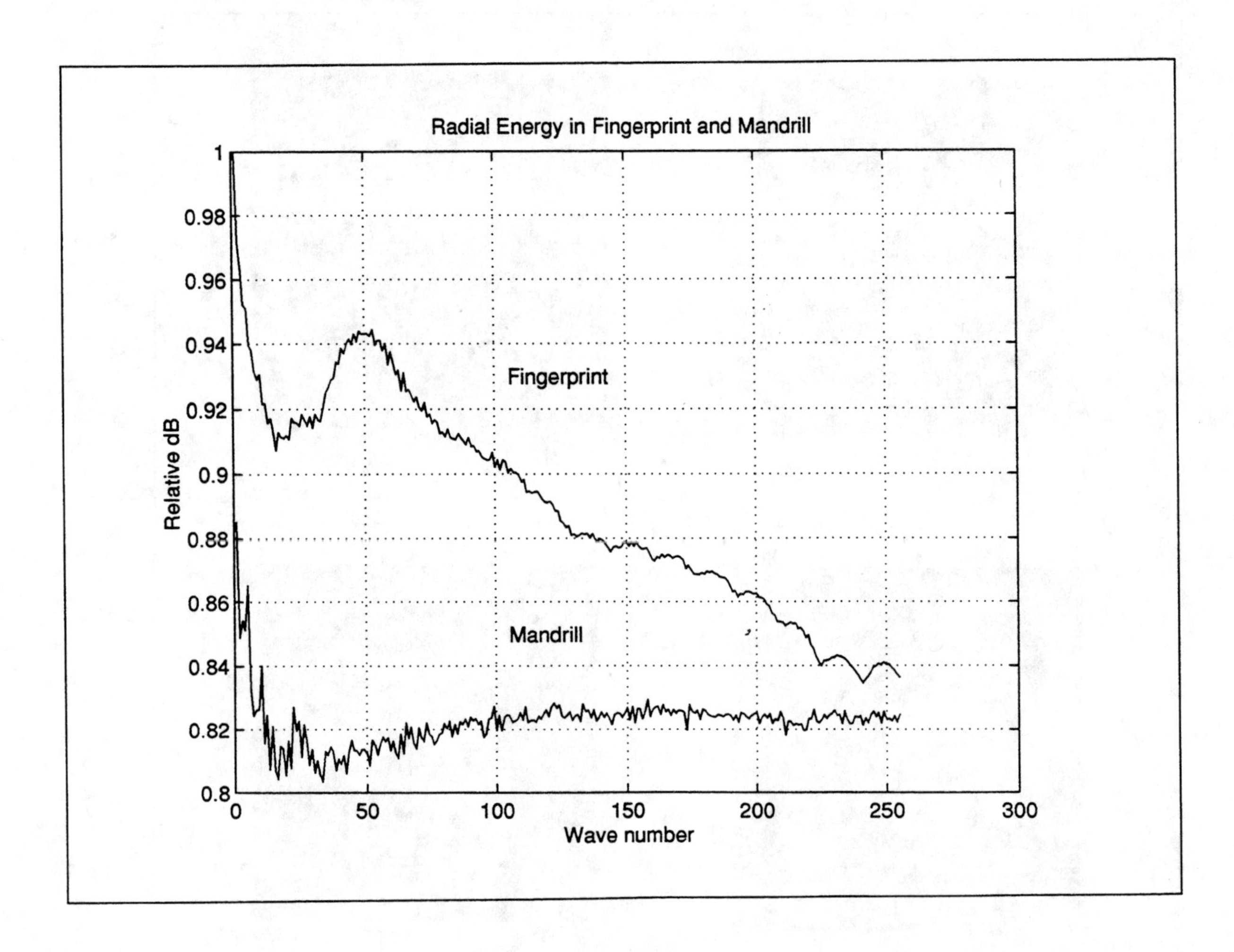

Figure 2. Average radial energy of 4000 NIST4 fingerprint images and the Mandrill.

2.2. Finding local direction in the frequency domain.

The 2D DFT can be used to find direction of ridge flow. Ridge direction is orthogonal to the direction of maximum frequency. Figure 3 shows a patch of ridges oriented at 0 and 37 degrees and the DC centered log normalized magnitude of their spectra. The peaks in the spectra show the direction of highest frequency which is orthogonal to the ridge flow.

The local window needs to be small enough so that the ridge structure is, piecewise, low curvature. The size of the DFT should be small to minimize the computational load but large enough to separate the low frequency noise (radii less than 30) from the frequency components that contribute to the ridge structure (between radii 30 and 70).

With a 64 point DFT the contributing frequencies would lie between radii 5 and 8 while the noise would lie between 1 and 4. This isolates the frequencies of interest and provides isolation from noise. A 64 x 64 pixel local window is too large to assure that most local window would be piecewise low curvature.

With a 16 point DFT the contributing frequencies would lie between radii 1 and 2. The noise components would be in the DC component and at radius 1; there is not enough isolation between the noise and the structural components. A 16 x 16 pixel local window would assure us that most of the window would contain low curvature ridges.

The best compromise is to use a 32 point 2D DFT on a 32 x 32 pixel local windows. Most of the local window content would be low curvature. There would be good isolation between noise and signal. The noise would be at radius 1 and the signal lies between radii 2 and 5.

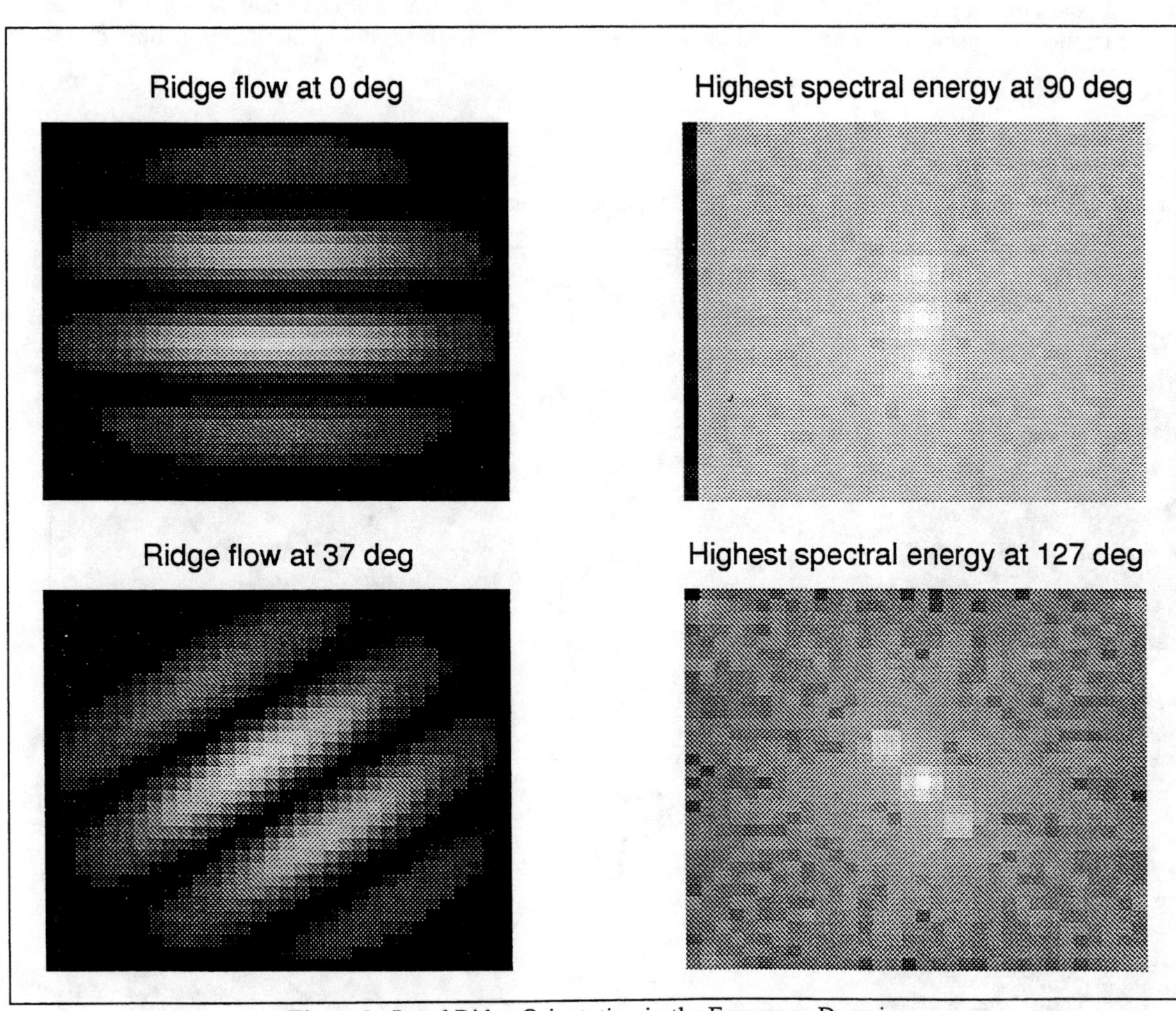

Figure 3. Local Ridge Orientation in the Frequency Domain.

2.3. Directional image enhancement

The enhancement process consists of finding and then amplifying the frequency components that contribute to the ridge structure (these are the ones with the most energy between radii 2 and 5). Just amplifying the frequency with the peak energy will distort the ridge structure. This is analogous to decomposing a square wave into a series of sine waves: amplifying only the frequency with the maximum energy would distort the time domain signal. There is more than one salient spectral value, reflecting the fact that more than one frequency contributes to the underlying ridge structure. The enhancement process finds the eight 3x3 areas with the highest energy and then amplifies all frequency components taking care not to amplify a component more that once. Computing the inverse 2D DFT of the modified spectrum results in an image with more dynamic range and improved signal to noise ratio.

2.4. EXAMPLES

The examples that follow illustrate the directional frequency domain enhancement in low curvature, core, delta, high curvature, cut, and creased areas. In all figures the 32 x 32 raw image, the peak and eight dominant directions as found in the frequency domain, and the enhanced image are shown. The dominant directions are normalized by the peak to show their relative strengths.

2.5. Low Curvature Regions

Figure 4 shows that for low curvature regions the dominant directions are consistent and the enhancement is accurate.

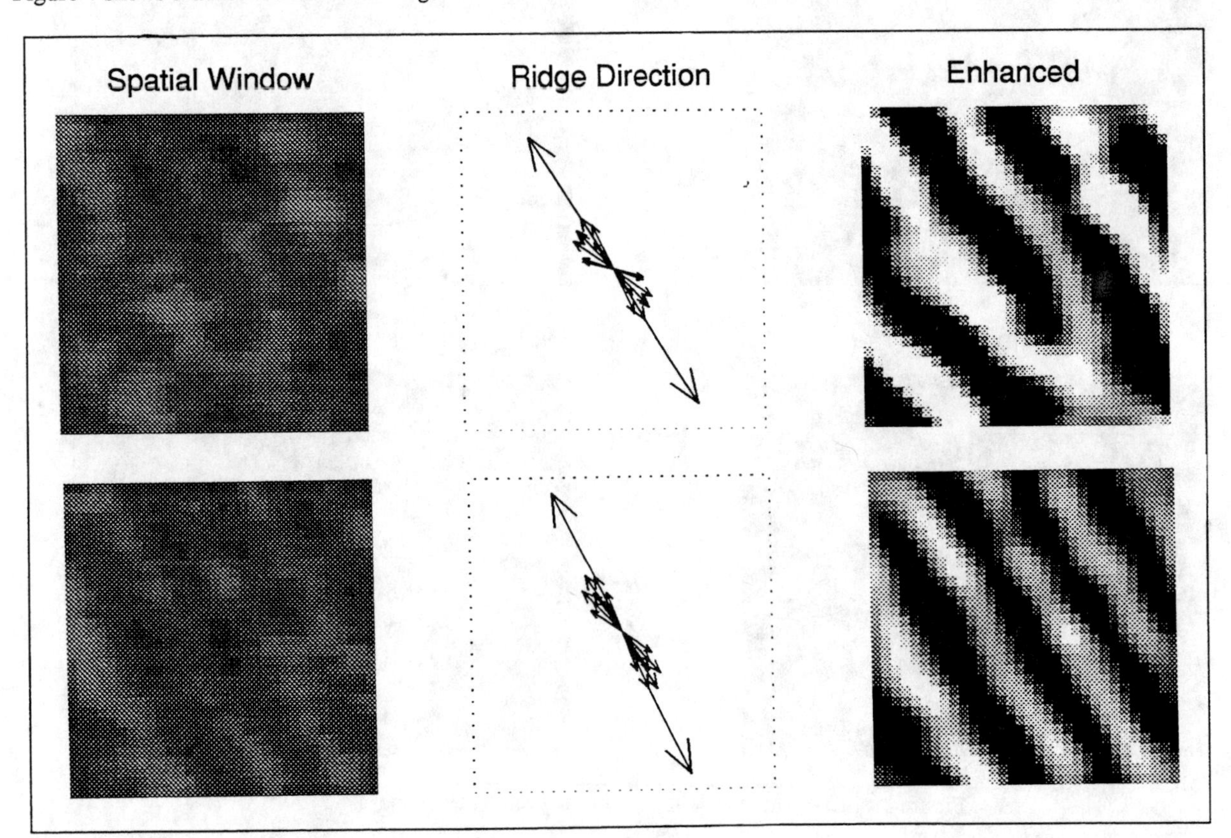

2.5.1. Cut and Creased Regions

Figure 5 shows the enhancement for cut and creased regions (scarred regions behave similarly). Spatial domain methods do poorly in these regions because pixel intensity along the artifact is high with little variation. The correlation along the artifact is higher than along any other direction. Unless the cut, scar, or crease is in the same direction as the ridge flow the incorrect direction is selected. The frequency domain method performs well in these areas because these artifacts have little effect on the frequency of the ridges.

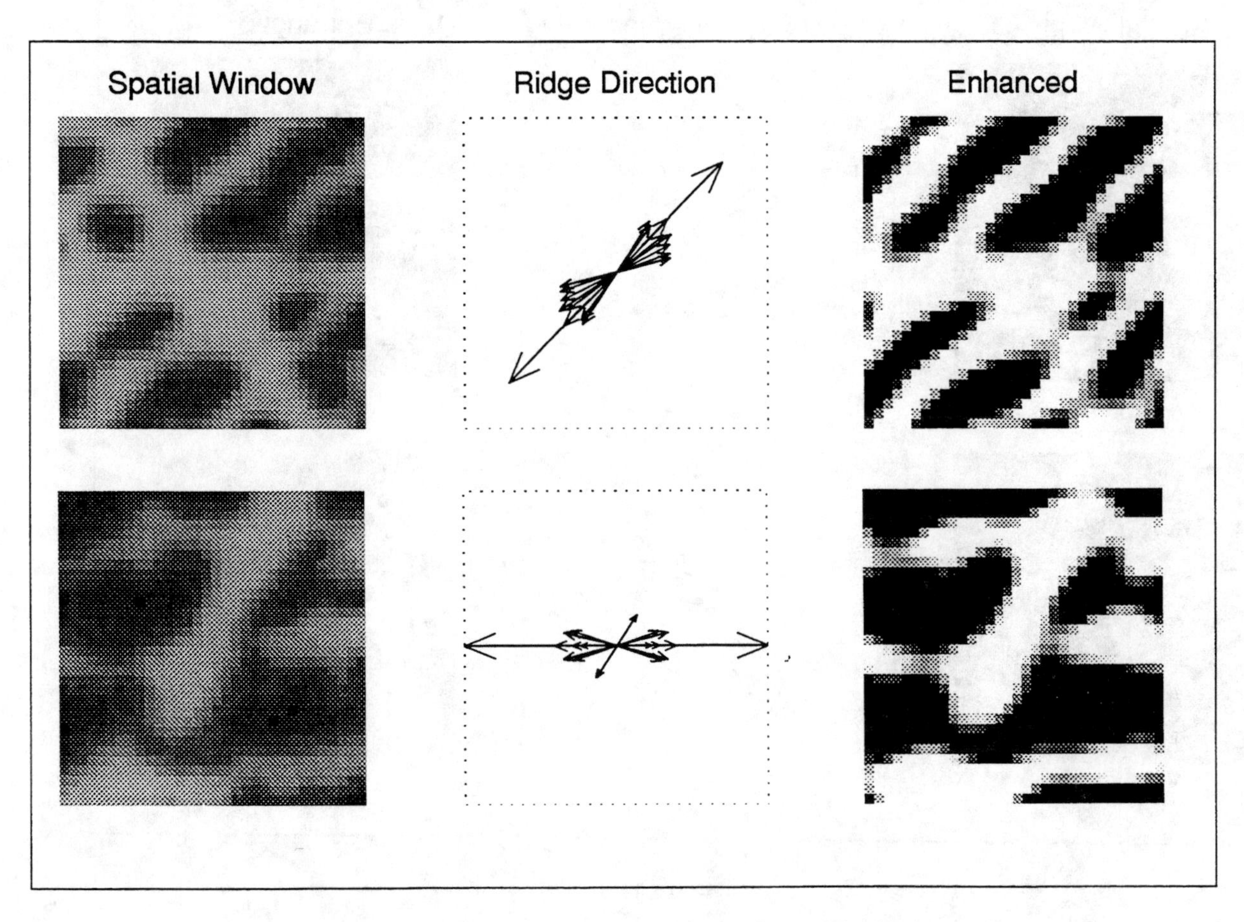

Figure 5. Enhancement in Cut and Creased Regions

2.5.2. High Curvature Regions

Figure 6 shows enhancement in high curvature areas. Spatial domain methods typical do poorly in these regions because they correlate with only one direction and tend to look like noisy areas with no dominant direction. The frequency domain method recognizes that more than one direction makes up the ridge structure and enhances all these directions.

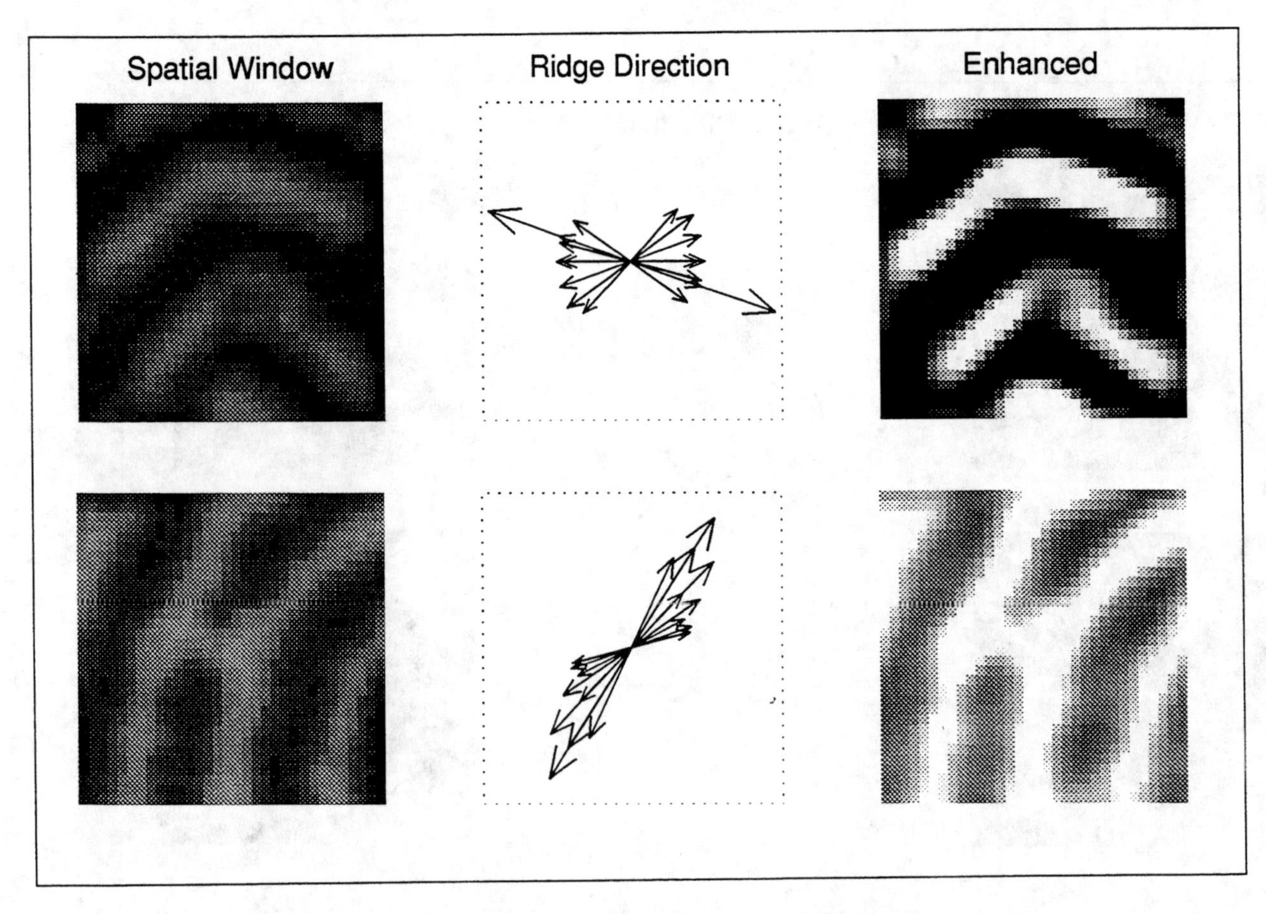

Figure 6. Enhancement in High Curvature Regions.

2.5.3. Core and Delta Regions

Figure 7 shows enhancement for core and delta regions. Again spatial domain techniques have a difficult time in these regions because there are multiple directions to the ridge flow. The frequency domain method performs well because it enhances more that a single direction.

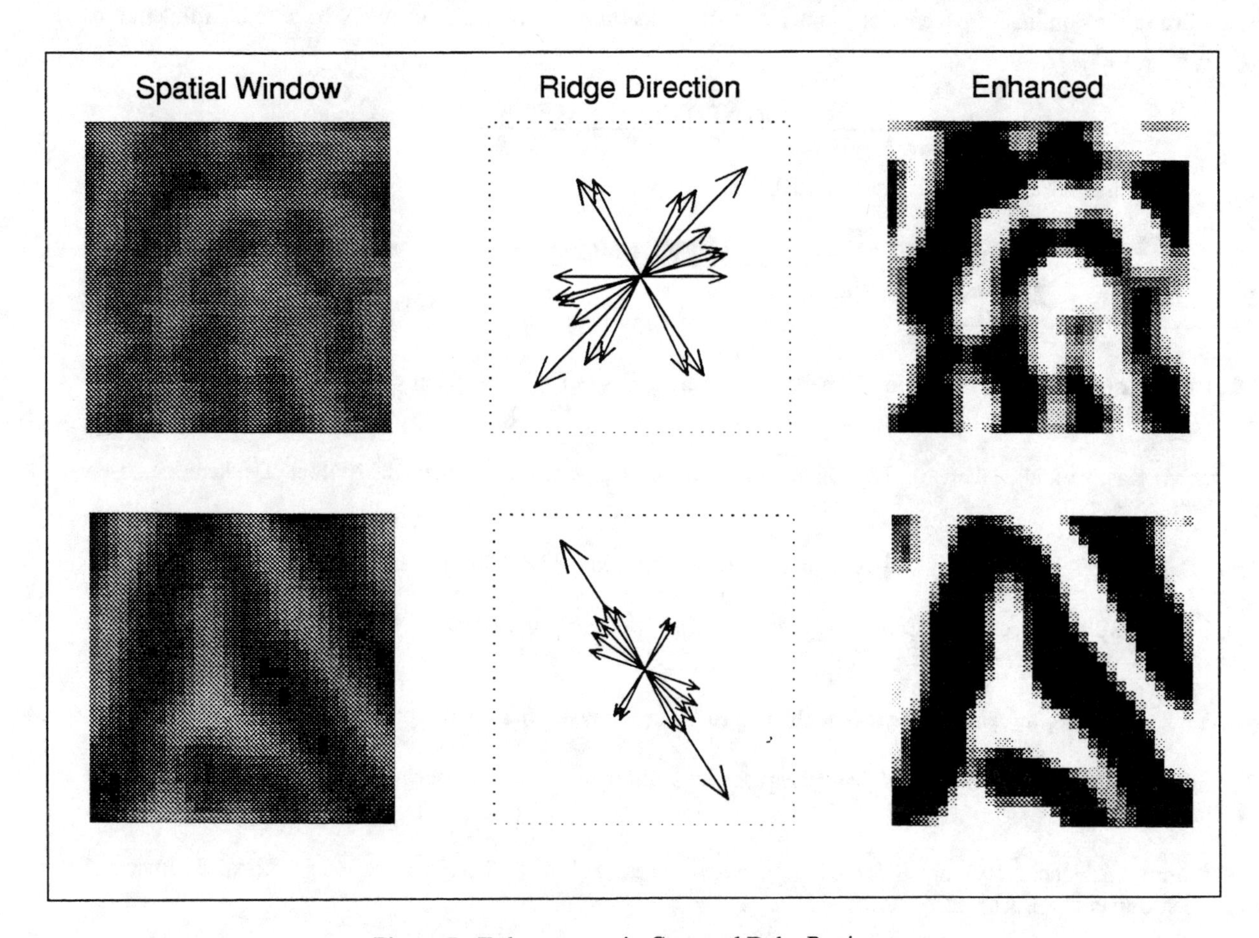

Figure 7. Enhancement in Core and Delta Regions.

3. CONCLUSIONS

The method presented here performs well in all areas of a fingerprints and performs better in noisy and high curvature regions where spatial methods fall short for two reasons.

1. The local direction estimate is not constrained to fit one of a set of pre-defined directions.

2. More than one direction is enhanced improving performance in high curvature regions.

3. The frequency domain method of finding local directions more robust in the presence of scars, creases, and cuts because they have little effect on the underlying frequency of the ridges.

Two major areas need to be addressed before this method makes it into an automated fingerprint identification system. The first is that the local enhanced images need to be "put back together" to form a single image. The most common way is to use 50% overlapped and windowed patches and then average the overlapping areas. The second is that the gain applied to the dominant frequency components must be determined. It will most likely have to be a function of the local window.

4. ACKNOWLEDGMENTS

This research was funded by Printrak International, Inc.

5. REFERENCES

---

[1] Michael Kass and Andrew Witkin, "Analyzing Oriented Patterns," Computer Vison, Graphics, and Image Processing, Vol. 32, pgs. 362-385, 1987.

[2] Bangachar Kasturi and Mohan M. Trivedi, *Image Analysis Applications*, chapter 10, Marcell Dekker, Inc., New York, 1990.

[3] Henry C. Lee and R.E. Gaensslen, *Advances in Fingerprint Technology*, CRC Press, Boca Raton, 1994.

[4] Clare D. McGillem, George R. Cooper, *Continuous and Discrete Signal and System Analysis*, chapter 3, Holt, Rinehart and Winston, Inc, New York, 1974.

[5] William K. Pratt, *Digital Image Processing*, chapter 10, John Wiley & Sons, New York, 1978.

[6] N. Ahmed and K.R. Rao, *Orthogonal Transforms for Digital Signal Processing*, chapters 5 and 6, Springer-Verlag, Heidelberg, 1975.

[7] J.W. Tarng, "Advanced Techniques for Digital Image Processing," M.S. Thesis, Dept. of Electrical Engineering, Texas Tech University, Lubbock, Texas.

[8] B.G. Sherlock, D.M. Monro, and K. Millard, "Fingerprint enhancement by directional Fourier filtering," IEE Proceedings-Vision, Image and Signal Processing, Vol. 141, No. 2, April 1994.

[9] C.I. Watson and C.L. Wilson, Nist Special Database 4, Fingerprint Database, National Institute of Standards and Technology, Advanced Systems Division, Image Recognition Group, March 17, 1992.

Fingerprint matching via spatial correlation with regional coherence

Al Ranalli

Printrak International
1250 N. Tustin Ave., Anaheim, Ca. 92807

ABSTRACT

A non-minutiae based technique for fingerprint matching is presented, based upon spatial correlation of fingerprint grayscale images. Normally, such techniques suffer from plastic distortion of the fingerprint ridges, which varies irreproducibly from one scan to another. This distortion can be seen as a process which reduces the degree of global spatial coherence of the fingerprint images. In the proposed approach, this limitation is accommodated by correlating small, windowed regions of a test image with the entire reference image. A score can then be ascertained, based on the partial correlations of these regions. The technique lends itself naturally to implementation by optical means.

Keywords: optical signal processing, fingerprint matching, matched filtering, joint transform correlation

1. INTRODUCTION

Most automated fingerprint identification system (AFIS) realizations evolve from a minutiae-matching paradigm, in which only the locations and orientations of fingerprint ridge endings and bifurcations (i.e. the minutiae) are considered relevant features for use in an algorithm-based matching scheme. Despite the compact representation of a minutiae file (a very desirable feature when considering digital storage and matching speed) and the relative success of such traditional AFIS systems, any effort to substantially improve matching accuracy cannot ignore the fact that there are fundamental limitations inherent in any technique, when used in isolation. For example, a procedure which automatically extracts minutiae from an electronic fingerprint image is prone to error, resulting in either missed or falsely registered minutiae - a high price to pay, considering the sparse-data nature of minutiae representation. As a result, sophisticated rules must be developed to attempt to accommodate the erroneous and overlooked minutiae. Nonetheless, there is still a basic uncertainty as to whether or not a given minutia is valid. Therefore, it is desirable to find an independent scheme for comparing two fingerprint images in a highly objective and robust manner, to complement minutiae-based methods.

Spatial correlation of the images, representing the degree of overlap between their ridge patterns, would appear to be an attractive scheme, were it not for its computational complexity and, more importantly, the detrimental consequences for the correlation operation due to plastic distortion of the fingerprint surface during image capture. The technique presented in this paper attempts to address this latter issue by proposing to overlap only small, individual regions of a fingerprint image (properly windowed) with a reference image, and deriving a match score, based on the collective overlapping of many such regions. While little can be done, outside of employing the fast Fourier-transform (FFT) algorithm, to improve the computational complexity of this scheme in a digital realization, most of the operations involved are easily performed in real time, via optical signal processing (Vander Lugt or Joint Transform correlation). Thus, our motivation for pursuing this research is the development of a hybrid optical processor for non-minutiae based matching.

This paper is organized as follows. Section 2 describes a Joint Transform optical spatial correlator, introducing terminology and definitions which are used throughout the paper. Section 3 describes a possible comparison scheme, and presents results of some of our studies. Section 4 discusses the potential role of the strain field, a by-product of the correlation scheme, in conditioning the decision process. Finally, section 5 concludes the paper with observations concerning the effectiveness of this technique and its efficient optical implementation.

2. OPTICAL SPATIAL CORRELATOR

The spatial Fourier-transforming property of ordinary lenses (for monochromatically illuminated objects) has led to numerous applications of optically-realized, two-dimensional spatial signal processing techniques, including holography[1,2]. For image matched-filtering applications, an optical architecture is sought for which the system is "tuned" to elicit a strong response when an input image strongly resembles a stored, comparison image. Borrowing some terminology from classical holography, the input and stored images will be referred to as object and reference images, respectively.

Fig. 1 shows one possible implementation of a hybrid optical-electronic spatial heterodyning system[3]. The object and reference images are coherently illuminated simultaneously at the front focal plane (FFP) of a lens, one focal length in front of the first-contacted lens surface. The CCD camera, located at the back focal plane (BFP) of the lens, records the power spectrum of the composite object/reference input image. By further Fourier-transforming this spectrum, an autocorrelation image is produced. This autocorrelation provides information regarding the self-similarity of the input image. Since the input image consists of both the object and reference images, non-superposed (hence the name: Joint Tranform), the existence of a large cross-correlation peak would suggest an occurrence of the reference image within the object image (or vice-versa). For example, Fig. 2 shows a cross-section of the autocorrelation image produced when the object image is the word "spatial" and the reference image is the letter "a". Note the correlation peaks at the locations of the occurrences of the reference "a" within the object image.

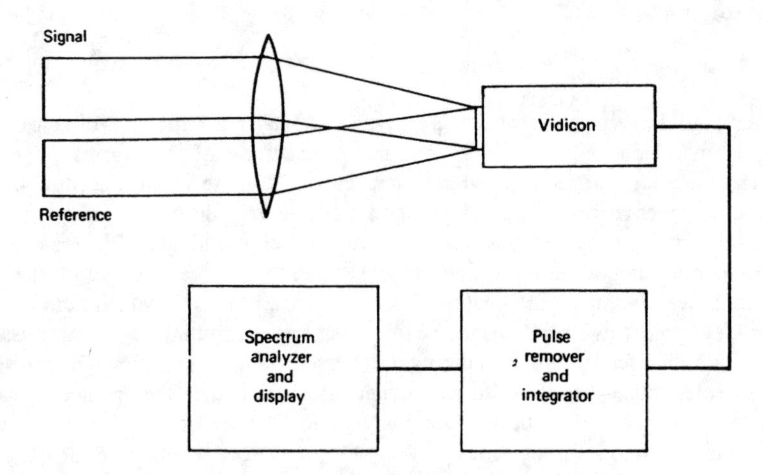

Fig. 1. A hybrid optical-electronic spatial heterodyning system[3].

Fig. 2. Correlation detection of the letter "a" using hybrid system[3].

Mathematically, let the input image, containing both the object and reference images, be designated by $\psi(x,y)$, where x and y are the Cartesian coordinates of the FFP. What appears at the BFP is the Fourier transform of the input image, $\Psi(u,v)$, where

$$\Psi(u,v) = \int\int \psi(x,y)\exp(-j2\pi[ux+vy])dxdy \tag{1}$$

where $j = \sqrt{-1}$, and u and v are the spatial frequency variables corresponding to the x and y dimensions, respectively. They are related to the physical dimensions, x_{BFP} and y_{BFP}, via the following relations:

$$u = \frac{x_{BFP}}{\lambda f} \tag{2a}$$

$$v = \frac{y_{BFP}}{\lambda f} \tag{2b}$$

where f is the lens focal length and λ is the wavelength of the (monochromatic) light illuminating the images. Since the CCD camera is only capable of measuring average intensity, it is the spatial power spectrum which is actually recorded. According to the Wiener-Khinchin theorem[4], this power spectrum is the Fourier transform of the autocorrelation sequence. Thus, the magnitude of the autocorrelation sequence can further be computed optically by simply placing the (coherently illuminated) CCD image at the FFP of another lens, and observing the pattern at its BFP.

Let the (non-overlapping) object and reference images be denoted by $\psi_O(x,y)$ and $\psi_R(x,y)$, respectively. The image representing the autocorrelation sequence is then given by

$$\phi(x,y) = \int\int |\Psi(u,v)|^2\exp(j2\pi[ux+vy])dudv \tag{3a}$$

$$= \int\int |\Psi_O(u,v) + \Psi_R(u,v)|^2\exp(j2\pi[ux+vy])dudv \tag{3b}$$

$$= \int\int (\, |\Psi_O(u,v)|^2 + |\Psi_R(u,v)|^2 + \Psi_O(u,v)\Psi^*_R(u,v) + \Psi^*_O(u,v)\Psi_R(u,v)\,)\exp(j2\pi[ux+vy])dudv \tag{3c}$$

where $\Psi_O(u,v)$ and $\Psi_R(u,v)$ represent the Fourier transforms of the object and reference images, respectively, and the asterisk denotes the complex conjugate. Since $\psi_O(x,y)$ and $\psi_R(x,y)$ are non-overlapping, the first two terms in equation (3c) will be zero for all points outside of either image. Thus, we can model the optical process numerically by restricting the domain exclusively to those regions, and computing only the last two terms in (3c). Note that these terms represent the inverse-Fourier transform of twice the real part of the product $\Psi_O(u,v)\Psi^*_R(u,v)$. Optically, an inverse-transform operation is identical to the transform operation, except for a reflection of the u and v axes about the origin. In the work which follows, this quantity will be computed for centered object and reference images, not adjacent ones. This simplification is entirely equivalent (except for a constant coordinate shift), but necessary, since numerical evaluation of the transforms becomes exponentially complex with increasing image size.

3. REGIONAL FINGERPRINT MATCHING

When considering the cross-correlation of two fingerprint images, a problem immediately presents itself. Strain in the coordinates of the fingerprint surface will prevent two fingerprint images from superposing throughout the extent of the fingerprint. Although considerable strain may result simply from optical aberrations in the imaging system of the scanner, the most common source of strain is the plastic distortion of the fingerprint surface during the capture process. Thus, while a particular orientation of one fingerprint image with respect to an image captured from the same finger, taken at a different time, may cause common ridges in one part of the print to align, the two images must be "nudged" in order to align ridges in another part of the print. An appropriate analogy for this phenomenon is to associate the fingerprint surface with a rubber jigsaw puzzle. Although stretching and twisting forces may appear to cause the surface as a whole to undergo severe distortion, each individual piece is only slightly perturbed. It is only over several pieces, along any given direction, that small stresses accumulate into significant distortions.

To illustrate an example of this strain effect on the correlation properties, consider the images displayed in Fig.'s 3 and 4. These images were digitized from inked impressions resulting from having rolled the same finger in two consecutive trials. They are 512 by 512 pixel, 8-bit grayscale images. 512 being an integer power of two, the images lend themselves readily to efficient spectral computation, via the FFT algorithm[5].

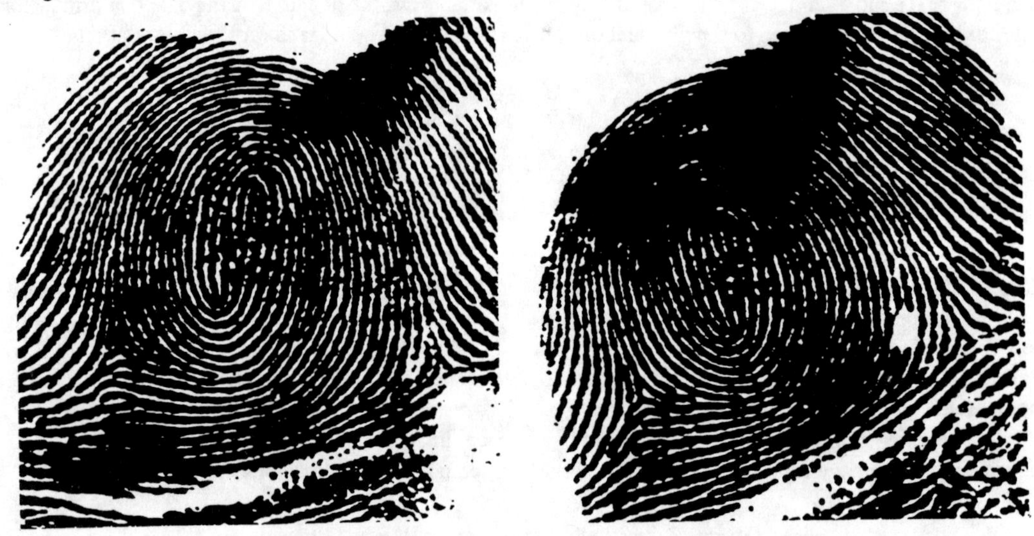

Fig. 3. Fig. 4.

Two rolled fingerprint impressions taken consecutively of the same finger.

Figure 5 shows the cross-correlation of the images in Fig.'s 3 and 4. Prior to the inverse-Fourier transformation described by equation (3c), the real part of the product $\Psi_O(u,v)\Psi^*_R(u,v)$ was multiplied by a Lorentzian band-pass weighting function, centered around 4 cycles/mm, with a (FWHM) bandwidth of about 2 cy/mm. Fig. 3 acts as $\psi_O(x,y)$, and Fig. 4 as $\psi_R(x,y)$. The bandpass filtering not only provides good noise suppression, but also helps to insure that the resulting correlation image is analytic. The correlation peak in Fig. 5 is only about 12% above the average value (the image has been contrast-enhanced for visibility). When the image in Fig. 6, which is an image of a completely different fingerprint, was correlated to that of Fig. 3, the result was a correlation peak which was 18% above the average, illustrating the uselessness of simply correlating fingerprint images blindly. Although Fig.'s 3 and 4 are from the same fingerprint, that print is locally stretched and twisted, so that any attempt to overlap them globally is futile.

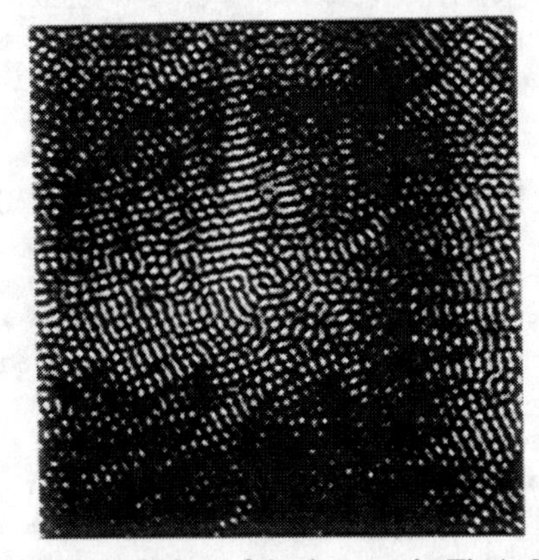

Fig. 5. Cross-correlation of the images in Fig.'s 3 and 4.

Fig. 6. An impression left by a different finger than that which produced Fig.'s 3 and 4.

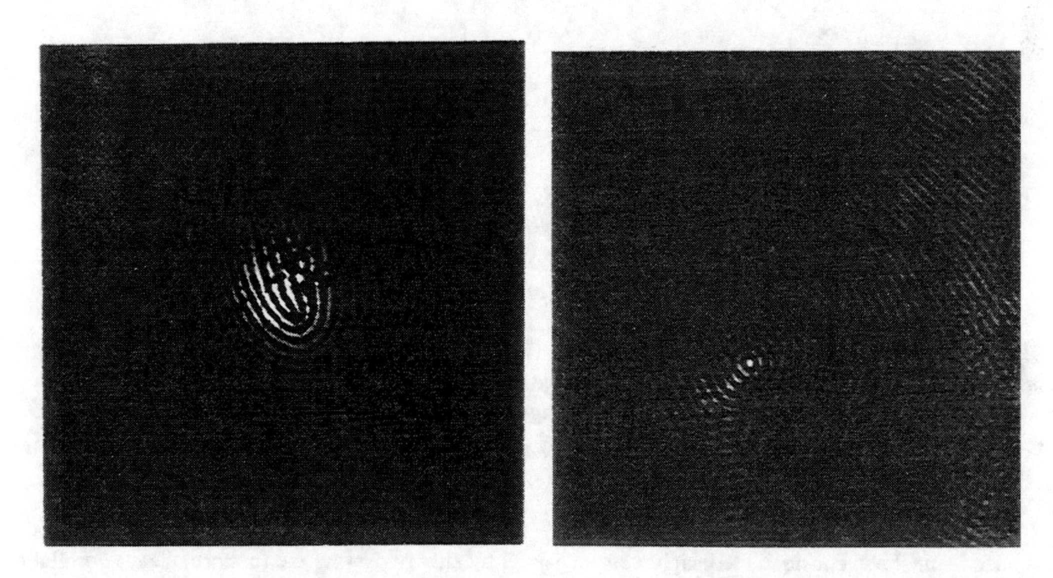

Fig. 7. Fig. 8.
A rotated, windowed region within Fig. 4, and the resulting cross-correlation with Fig. 3.

Figure 7 shows a portion of Fig. 4, windowed (again, to insure analyticity), rotated, and centered within a 512 x 512 image. When this fingerprint region is correlated against Fig. 3, the result is shown in Fig. 8, whose correlation peak is now more than 8 times larger than the average. Correlating rotated versions of the same fingerprint region with Fig. 3 shows that the correlation peak has a half-width of just under 5 degrees for the region width illustrated in Fig. 7.

As an example of a simple matching scheme, 24 points were chosen as region centers near the center of Fig. 4, and 5 rotations of each segment were correlated against Fig.3. The result peak values for the correlation images are shown in Fig. 9. The same operation was performed, sampling the (non-matched) fingerprint shown in Fig. 6. Those results appear in Fig. 10.

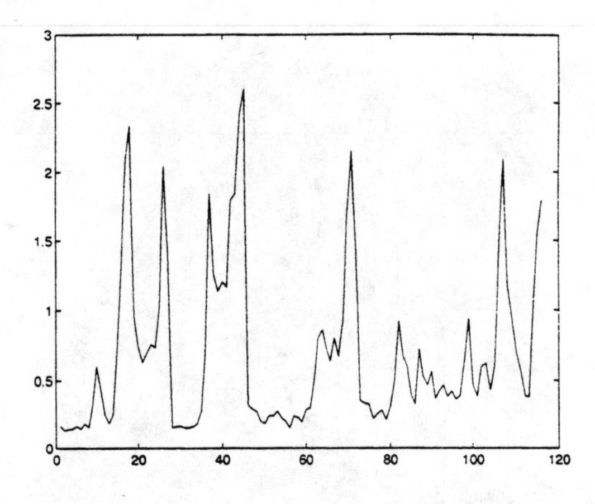

Fig. 9.
Peak correlation values comparing Fig. 3 with regions within Fig. 4. The horizontal axis represents the sample number, and the vertical axis represents correlation strength, in arbitrary units.

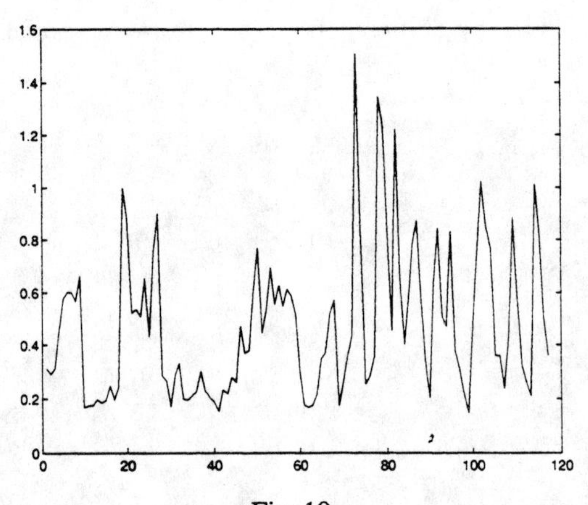

Fig. 10.
Peak correlation values comparing Fig. 3 with regions within Fig. 6

Although a comparison of two arbitrary fingerprint images necessarily requires that the images be normalized somehow, considerable insight can be gained by simply passing the (unnormalized) correlation peak values through a non-linearity, to accentuate strong correlations, and summing the outputs. To that end, the correlation peak values were used as exponents (base 10). The values, 10 raised to the power ϕ_{max}, were summed over all the correlation maxima, ϕ_{max}, yielding 103.7 for Fig. 9, and 0.064 for Fig. 10. Similar computations, performed for several other non-matching fingerprint images captured from the same scanning device, yielded sums which were consistently below 103, typically less than 1.

4. STRAIN FIELD CONSIDERATIONS

As noted in the previous section, a robust comparison of two image files necessarily requires some sort of normalization of the images - in general, not a trivial process. To be truly effective however, a comparison must also respect the global topology of the fingerprint surfaces. Adjacent regions in an object image must correlate to adjacent regions in the reference image, if the two images are truly matched. This information is not considered in the simple matching scheme outlined in section 3, even for normalized images.

Although our detailed matching algorithm is proprietary, and hence will not be disclosed here, it can be mentioned that topological inferences could be derived from the strain field resulting from the numerous correlations. Correlation peaks for a given object region are associated with (x,y) translations, and rotations. Thus, for every region there is a three-dimensional (strain) vector, the collection of which over all such regions forms a strain field. This field gives a conformal mapping of the object coordinate system to the image coordinate system, and can be used to compute conditional decision variables in a maximum-likelihood detection scheme. Two prints which are truly matched should lead to a strain field which is well-behaved. Used in this way, the strain field provides an excellent filter for hypothesis testing. It should be noted, however, that in order to get good resolution on this strain field, numerous correlations are required. Even for a hybrid semi-optical processor, many object regions must be isolated and used to modulate input images, and many correlation images must be read out and processed.

5. CONCLUSION

Spatial correlation of object regions with reference images provides a powerful tool for automated fingerprint image comparison, provided the object regions are of a useful size, and adequate spatial and rotational resolutions are respected. These requirements lead to the conclusion that a hybrid optical/electronic device is currently the only viable hardware for performing the necessary computations in a reasonable amount of time (e.g. 0.1 second per 512 x 512 grayscale image pair). The situation is exacerbated if topological inferences are to be used in hypothesis testing, since the strain field must be resolved to a sufficient degree to warrant a determination of whether or not it is well-behaved. Thus, image input/output becomes the processing bottleneck.

Continuing work in this area is concerned with the development of an adaptive comparison scheme, in order to reduce the number of rotations and/or the area of the spatial light modulators needed to properly represent the object and autocorrelation images, and to allow for an early-out rejection option. Considerations of the feedback variables used in such a scheme will drive the engineering of specialized hybrid hardware (possibly including servo-rotation and translation stages) for its effective implementation.

Also, spatial correlation may be used to verify hypotheses posed by minutiae-based, regional matching. By clustering minutiae associated with an object print, and finding the orientation of that cluster which yields the best match within the reference print, an estimate of the orientation angle for that region is established. The rotational dimension then being eliminated as a search variable, substantially fewer correlation operations are required.

6. REFERENCES

1. A. Papoulis, *Systems and Transforms with Applications in Optics*, McGraw-Hill, N.Y., 1968.
2. R. J. Collier, C. B. Burckhardt and L. H. Lin, *Optical Holography*, Academic Press, N.Y., 1971.
3. W. T. Cathey, *Optical Information Processing and Holography*, Wiley, N.Y., 1974.
4. A. Papoulis, *Probability, Random Variables, and Stochastic Processes*, McGraw-Hill, N.Y., 1984.
5. J. W. Cooley, J. W. Tukey, "An algorithm for the machine calculation of complex Fourier series", *Math. Comput.*, Vol. 19, pp. 297-301, 1965.

Fingerprint sensor using fiber optic faceplate

Masahiro Shikai, Hajime Nakajima, Toshiro Nakashima and Kazuo Takashima

Mitsubishi Electric Corp., Industrial Electronics and Systems Laboratory
Amagasaki, Hyogo 661, Japan

ABSTRACT

This paper presents a new fingerprint sensor for automated fingerprint identification systems. This fingerprint sensor consists of a light source, a charge coupled device and fiber optic faceplates. Because of the fiber optic faceplates, this fingerprint sensor does not have the space for image formation that a lens has and thus can be miniaturized. The size of the prototype fingerprint sensor, which has a 12 mm × 18 mm fingerprint input face, is 48 mm × 72 mm × 44 mm. The verification performance is evaluated by using this prototype.

Keywords: fingerprint sensor, fiber optic faceplate, fingerprint identification, security

1. INTRODUCTION

Fingerprints have been an effective means of identifying individuals for a long time. Based upon a century of examination, it has been verified that the chance of two people having the same fingerprint is less than one in a billion.[1] In recent years, automated fingerprint identification systems (AFISs) have been developed for not only criminal investigation but also for physical and virtual access control.[2-4] Typical examples of access control using AFISs are access control to facilities, data access to computers, and automated teller machines.

The AFISs for access control need fingerprint sensors (also called "live scans") that can capture high-resolution images of fingerprints directly and rapidly. Recently, the miniaturization of fingerprint sensors for desk top use, access control to private houses, personal digital assistants and so on has been demanded. Fingerprint sensors using optical methods have been developed mainly. The prism total internal reflection (prism TIR) method is one of the most popular of these.[4] A fingerprint sensor using the prism TIR method consists of a light source, a prism and a camera with a lens. However, it is difficult to reduce the size of the whole sensor because the prism TIR method needs not only optical parts but also a space for image formation. Further, the prism TIR method introduces trapezoidal distortion and a partial focal shift in the fingerprint image because the camera faces the input face of the prism obliquely.

In this paper, we introduce a new optical fingerprint sensor using fiber optic faceplate (FOF). This new sensor solves the above problems. The basic principle of this sensor, its optical design and the experimental results of the prototype will be described.

2. BASIC OPERATING PRINCIPLE

Figure 1 shows the basic structure of the fingerprint sensor using an FOF. The FOF is a block formed of a bundle of optical fibers having two end faces which transmits an image from one end face to the other. The resolution corresponds to the diameter of each optical fiber of the FOF. The basic structure consists of light emitting diodes (LEDs), a slanted FOF and a charge-coupled device (CCD). The CCD is directly connected to the non-slanted end face of the FOF. The slanted end face of the FOF is used as the fingerprint input face. When the LEDs illuminates the fingerprint input face across the optical fibers of the FOF, a fingerprint image is generated there and is transmitted direct to the CCD by the FOF.

Figure 2 shows the principle of generating a fingerprint image. The FOF consists of many optical fibers, in this figure, however, we show only one optical fiber to avoid complexity. The total internal reflection on the fingerprint input face is used for this method as in the prism TIR method. All of the light beam propagated in the optical fiber core is required to be completely reflected on the slanted face when there is no finger present for the sake of the high contrast ratio of the fingerprint image. Therefore the slanted fingerprint input face is required to satisfy the following condition.

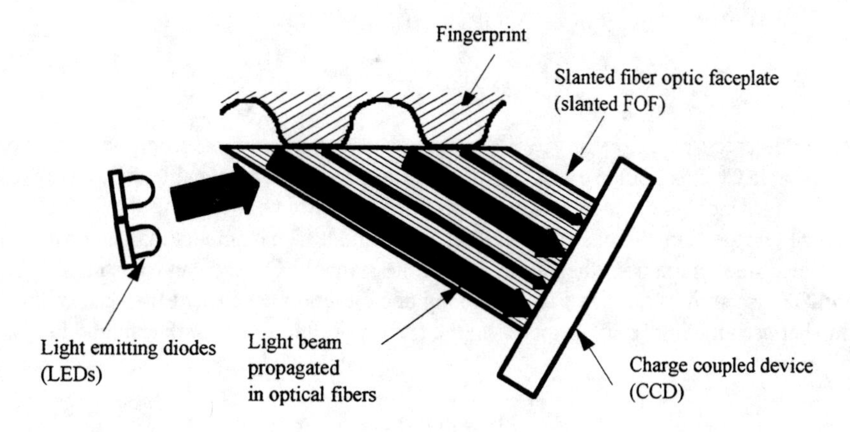

Fig. 1 Basic structure of a fingerprint sensor using FOF.

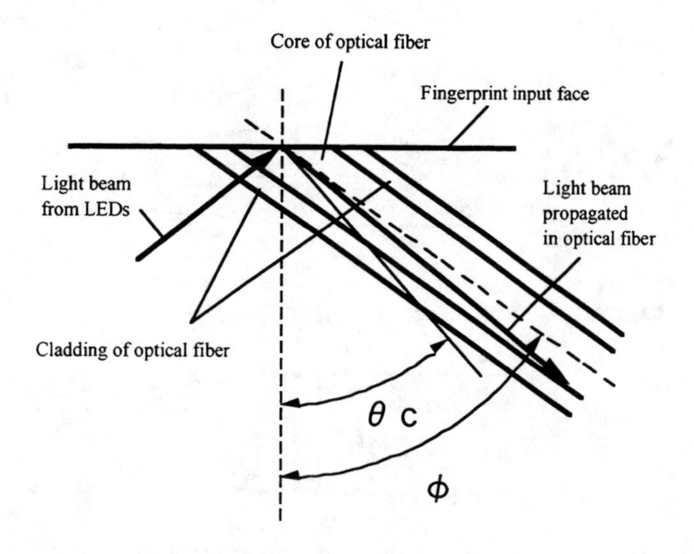

Fig. 2 Principle of generating a fingerprint image.

$$\theta c + \sin^{-1}(NA) < \phi, \tag{1}$$

where ϕ is the angle from the normal line of the fingerprint input face to the orientation of the optical fibers, θc is the critical angle at the interface between air and the optical fiber core, and NA is the numerical aperture of the optical fiber. The light beam from the LEDs is totaly reflected at the fingerprint input face with no finger present and is coupled to the optical fibers. When the ridges of a fingerprint touch part of the input face, the total reflection condition is no longer satisfied at that particular location. In this way, a fingerprint image which is dark at the ridges of a fingerprint and bright at the valleys is formed on the fingerprint input face. The fingerprint image is then directly transmitted to the CCD by the FOF.

This sensor has the following features.

- A space for image formation is not required, because of the FOF. Thus the size of the sensor can be smaller than sensors using lenses.
- The fingerprint image of this sensor is free of trapezoidal distortion and a partial focal shift, while there are both in the image produced by the conventional prism TIR method.

3. OPTICAL DESIGN

3.1. Shrinkage of image

For higher verification accuracy, more fingerprint information must be captured from a wide area of the finger. However, the imaging area of a usable CCD is limited by its package size and cost. The reasonable imaging area selected for the prototype is 6.6 mm × 5.0 mm (1/2 inch type). Therefore, a tapered FOF is inserted between the slanted FOF and the CCD, as shown in Fig. 3. In the tapered FOF, the diameter of an optical fiber on its output face is smaller than on its input face. Thus the tapered FOF can shrink the size of the image from the output face of the slanted FOF and can transmit a shrunk image to the CCD. Further, the slanted FOF also shrinks the size of image along one direction because of the slant of the fingerprint input face. Thus the relationship between the image size Lx × Ly on the fingerprint input face of the slanted FOF and the image size lx × ly on the CCD is given as

$$lx = m\, Lx, \tag{2}$$
$$ly = m\, Ly \sin(\pi / 2 - \phi), \tag{3}$$

where m is the magnification of the tapered FOF.

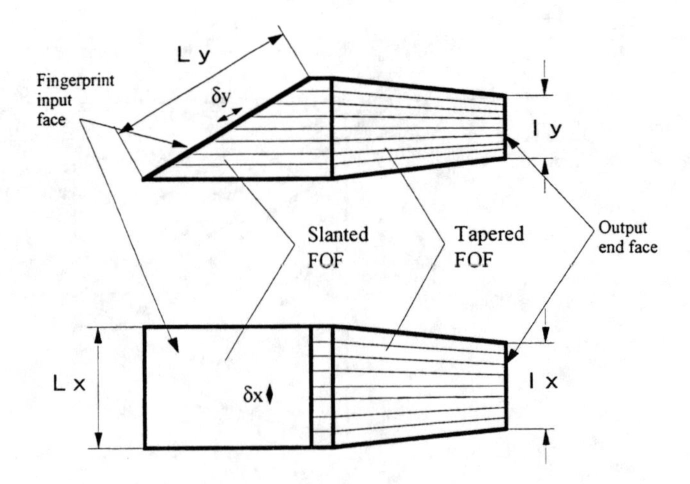

Fig. 3 Shrinkage of fingerprint image.

3.2. Resolution of FOFs

The sensor must be designed so that its resolution is sufficiently smaller than the spacing of a fingerprint. The resolution depends on the diameter of each optical fiber of the FOFs. In consideration of the slant of the fingerprint input face, the resolutions δx and δy of the image on the fingerprint input face are given as

$$\delta x = D, \tag{4}$$
$$\delta y = D / \sin(\pi / 2 - \phi), \tag{5}$$

where D is the diameter of the optical fiber of the slanted FOF. Because the spacing between ridges of a fingerprint is approximately 500 μm, we determined that D be 25 μm, which is about one twentieth of this spacing. Further, in the tapered FOF, the diameter selected for the optical fiber on the input face is also 25 μm to maintain the resolution of the fingerprint image. Figure 4 shows a part of the fingerprint image on the output end face of the slanted FOF. The small hexagon patterns shown in the image are the ends of the optical fibers. About ten optical fibers exist over the width of the ridge, which the dark part of the image corresponds to.

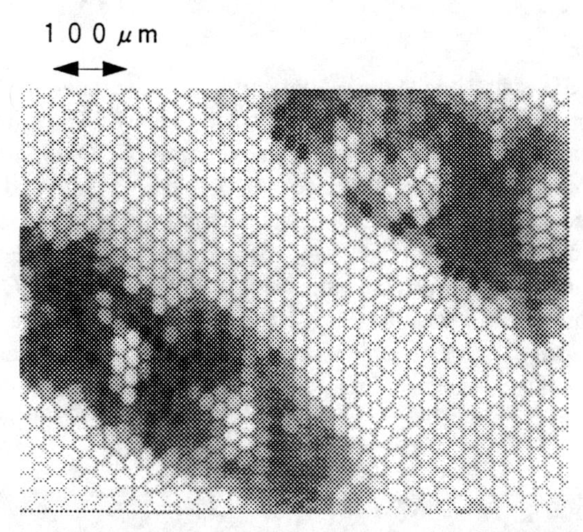

Fig. 4 Output end face of slanted FOF with a fingerprint.

3.3 Elimination of scattered light

While propagating from the LEDs to the fingerprint input face, a part of a light beam is scattered in the slanted FOF, because the FOF has a complex structure which consists of many optical fibers. The scattered light doesn't remain in the core of each optical fiber and uniformly illuminates the whole of the output face of the slanted FOF, as shown in Fig. 5(a). The uniform illumination causes an offset which reduces the contrast ratio of the fingerprint image. To eliminate the scattered light, the tapered FOF has optically absorbent material between the optical fibers, as shown in Fig. 5(b). Only the scattered light, which crosses the optical fibers, is absorbed by the optically absorbent material in the tapered FOF. While the imaging light beam propagated in the optical fiber core never comes into contact with the optically absorbent material between the optical fibers and reaches the end of the tapered FOF without loss. It was experimentally observed that a tapered FOF with the optically absorbent material improved the contrast ratio of the fingerprint image by approximately 20% relative to a tapered FOF without the optically absorbent material. The value of the contrast ratio is the ratio of the brightness at valleys to the brightness at ridges in a fingerprint image.

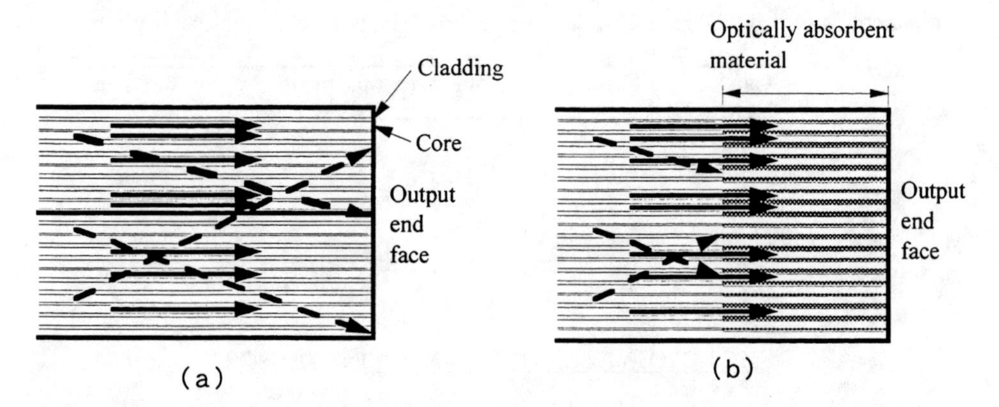

Fig. 5 Elimination of scattered light with optically absorbent material.
(solid line : light beam propagated in each optical fiber core, broken line : scattered light)

3.4. Gap between the FOF and the CCD

If the imaging plane of the CCD is attached to the output end face of the tapered FOF without a gap, a moiré pattern is generated which is superimposed on the fingerprint image, as shown in Fig. 6(a). The moiré pattern is produced by interaction between the pattern of the fibers and the arrangement of image pixels of the CCD. In image processing, such a moiré pattern introduces errors, and therefore should be eliminated. Placing a gap between the FOF and the CCD is effective in reducing the moiré pattern, while the gap also reduces the contrast ratio of the required fingerprint image. Table 1 shows the experimental results indicating the contrast ratio of the fingerprint image and the presence of the moiré pattern according to the gap. The gap is produced by inserting a thin spacing glass between the output end face of the tapered FOF and the imaging plane of the CCD. The value of the gap in Table 1 shows the thickness of the inserted spacing glass of which refractive index is 1.52. As shown in Table 1 and Fig. 6(b), the moiré pattern isn't generated if the gap is equal to or is bigger than 100 μm. On the other hand, as the gap increases, the image contrast of the fingerprint image decreases. Hence a minimum gap value which produces no moiré pattern should be selected.

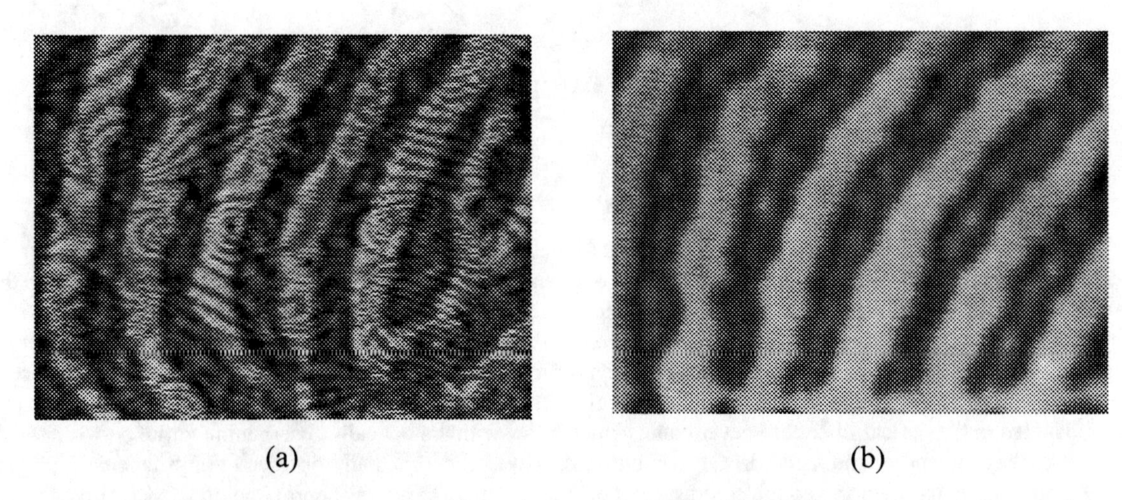

Fig. 6 Fingerprint images versus the gap between the FOF and the CCD: (a) the gap is 0 μm and (b) the gap is 100 μm.

Table 1 Contrast Ratio of Fingerprint Image and the Presence of Moire Pattern versus the Gap between the FOF and the CCD

Gap	Contrast Ratio of Fingerprint Image	Moire Pattern
0 μm	4.44	clear
100 μm	2.70	none
200 μm	2.64	none
300 μm	2.27	none

4. PROTOTYPE OF THE FINGERPRINT SENSOR

Table 2 shows the specifications of the prototype fingerprint sensor using FOFs. Ten LEDs illuminate the fingerprint input face of the slanted FOF uniformly. The FOFs, the spacing glass and the CCD are bonded with transparent adhesives.

Figure 7 shows the appearance of the prototype fingerprint sensor with a driving circuit for the CCD. The prototype fingerprint sensor outputs fingerprint images as a video signal. The case size is 48 mm × 72 mm × 44 mm.

Table 2 Specifications of the Prototype Fingerprint Sensor Using FOFs

LEDs	Number	10
Slanted FOF	Size of fingerprint input face: Lx × Ly Angle of fingerprint input face: ϕ Diameter of optical fiber: D Optically absorbent material	12 mm × 18 mm 60° 25 μm none
Tapered FOF	Magnification: m Diameter of optical fiber on the input end face Optically absorbent material	0.55 25 μm contained
Spacer glass	Thickness	100 μm
CCD	Size of imaging area	6.6 mm × 5.0 mm

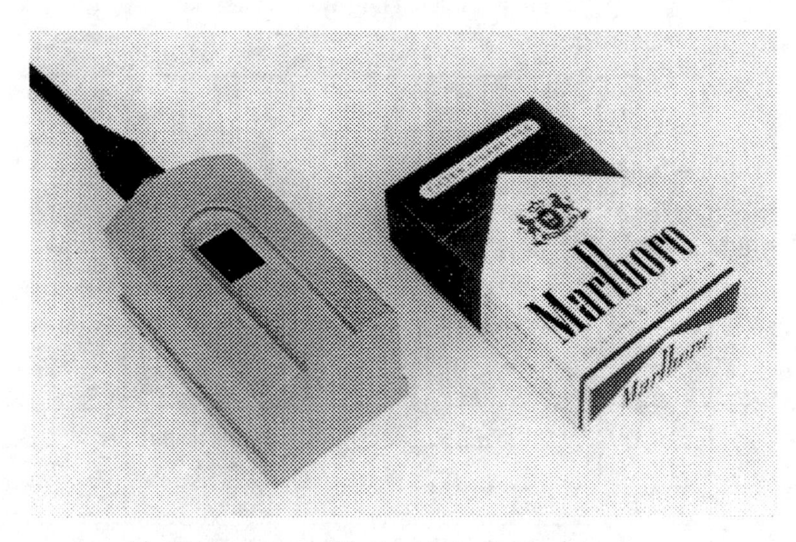

Fig. 7 Appearance of the prototype fingerprint sensor.

Figure 8 shows the output images of the prototype fingerprint sensor. The dark part of the image corresponds to the ridges of the fingerprint. There is no trapezoidal distortion and no partial focal shift in these images. Figure 8(a) shows an image when the skin of the finger has sufficient oil or sweat and the valley of the fingerprint isn't shallow. The ridge pattern appears clearly. On the other hand, Fig. 8(b) shows an image when the skin of the finger has too little oil or sweat. The ridge pattern doesn't appear clearly and isn't stable. The quality of the fingerprint images depends on the condition of the fingers.

Figure 9 shows the result of the verification test of the prototype sensor and our fingerprint verification algorithm.[2] This fingerprint verification algorithm carries out the matching of minutiae. Minutiae are ridge ends and bifurcations in a fingerprint. In Fig. 9, the horizontal axis is the threshold of the matching degree of minutiae and the vertical axis is the error rate. In this verification test, the number of sample fingerprints is 960 (120 fingers × 8 times). When the threshold is 35%, the false accept error and the false reject error are 1% and 5% respectively. This performance achieves practical level although there is room for improvement. The causes of verification errors are mainly poor quality fingerprints (too little oil or sweat, shallow valleys) and position shifts of the fingers. If the adhesion of the fingerprint input face to skin were improved and the size of the fingerprint input face were enlarged, the verification performance would be better.

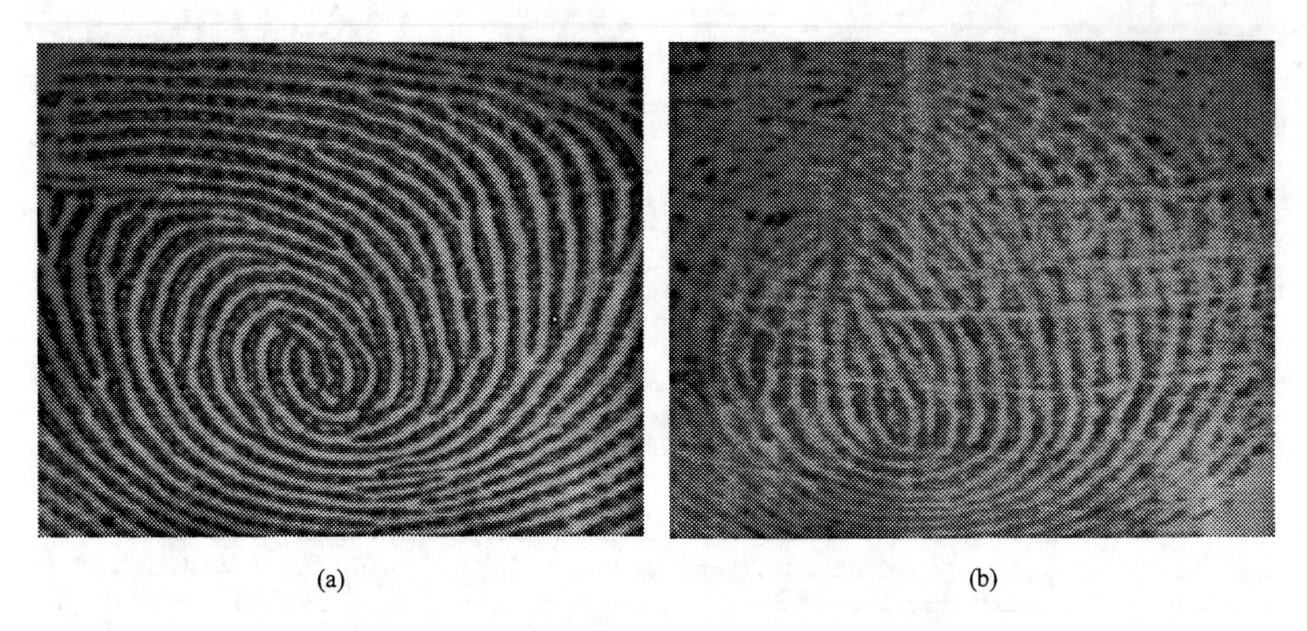

Fig. 8 Captured fingerprint images.

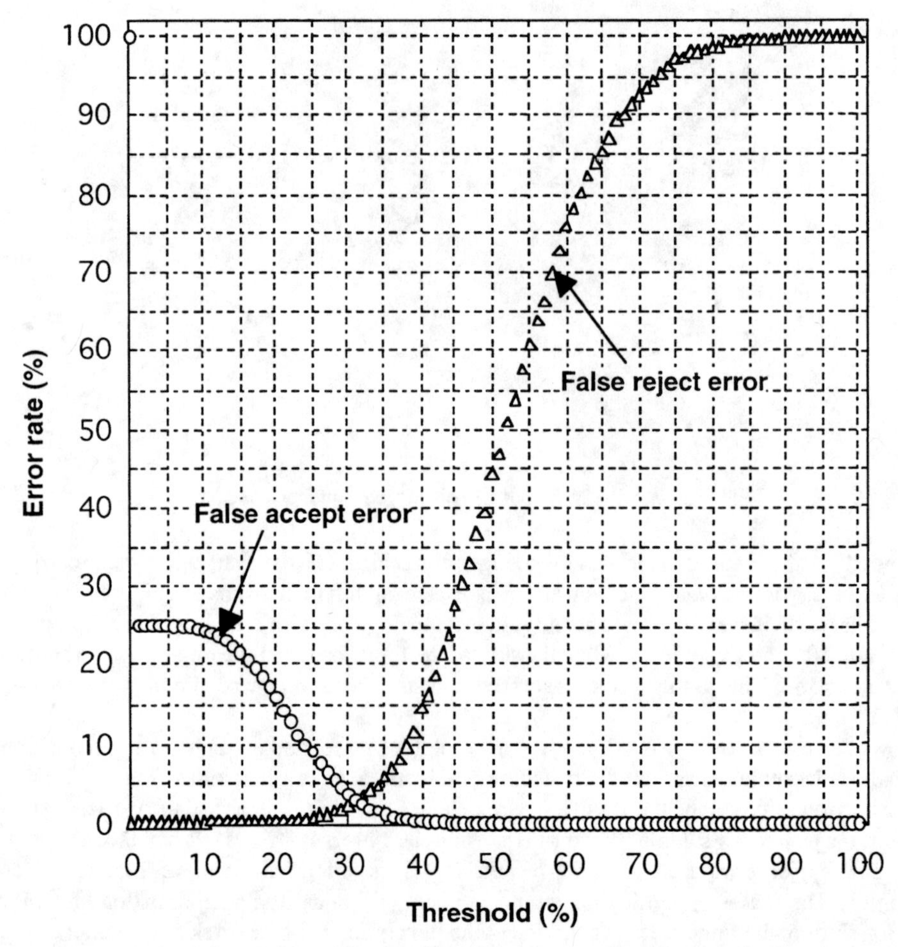

Fig. 9 Result of verification test using the prototype fungerprint sensor.

5. CONCLUSION

We have presented a new fingerprint sensor using FOFs for AFISs. A prototype fingerprint sensor has been designed and produced. Because it uses FOFs, the prototype fingerprint sensor is compact, and the fingerprint image is free of trapezoidal distortion and a partial focal shift. Further, the results of the verification test with the prototype show that this sensor is useful for AFISs.

6. ACKNOWLEDGMENT

We would like to express our appreciation to K. Sasakawa and H. Fujiwara for their help and cooperation and to I. Nakahori and T.Usami for their helpful advices.

7. REFERENCES

1. B. Miller, "Vital signs of identity," *IEEE Spectrum*, pp. 22-30, February 1994.
2. K. Sasakawa, F. Isogai and S.I kebata, "Personal verification system with high tolerance of poor quality fingerprints," *Proc. SPIE*, Vol. 1386, pp. 265-272, 1990.
3. S. Ozaki, T. Matsumoto and H. Imai, " A holder verification protocol using fingerprint," *Korea-Japan Joint Workshop on Information Security and Cryptology*, JW-ISC'93, 1993.
4. K. Morita and K. Asai, "Automated fingerprint identification terminals for personal verification," *NEC Res. & Develop.*, No. 83, pp.15-21, October 1986.

FaceIt: face recognition from static and live video for law enforcement[1]

Joseph J. Atick, Paul A. Griffin and A. Norman Redlich
Computational Neuroscience Laboratory
The Rockefeller University
1230 York Ave, New York, NY 10021, USA

and

Visionics Corporation
7-11 Green Street, Suite 15B, Metuchen NJ 08840,USA
http://www.faceit.com

ABSTRACT

Recent advances in image and pattern recognition technology -- especially face recognition – are leading to the development of a new generation of information systems of great value to the law enforcement community. With these systems it is now possible to pool and manage vast amounts of biometric intelligence such as face and finger print records and conduct computerized searches on them. We review one of the enabling technologies underlying these systems -- the FaceIt face recognition engine -- and discuss three applications that illustrate its benefits as a problem-solving technology and an efficient and cost effective investigative tool.

Keywords: face recognition, biometric systems, database search engine, surveillance, finger print, human face, identification, authentication.

1. THE NEED FOR BIOMETRIC DATA MANAGEMENT

The information age has lived up to its promise, "information at everyone's fingertips." Databases around the world are swelling with terabits of text, images and multimedia clips on every conceivable topic and are growing at astonishing rates. The internet has helped accelerate this growth and made the information stockpiles accessible to a wider community. However, the information age has not yet achieved its fullest potential since tools to manage the huge amounts of information and to conduct computerized searches have lagged behind. As anyone who has surfed the internet knows, information is as good as your ability to find it and organize it in a way that suits your purposes.

The situation is not any different when it comes to information resources within the law enforcement community. Large numbers of electronic databases containing biometric and other information are currently under development at various agencies at the Federal, state and local levels. Examples include databases containing facial photographs, finger print images and other textual information on gang members, drug traffickers, terrorists and other criminal individuals. In addition DMVs across the country

---

[1] Some of the R&D work in this report was funded in part by a grant from the Office of Naval Research.

are in the process of modernizing their records by converting into digital form containing facial photographs and soon many will add finger print images as well.

Whether these databases are to be used for investigative purposes or to combat fraud, their utility will only be fully realized if cost-effective technology for efficient searching of records by biometric content is available to the law enforcement community. This is a challenge to the pattern recognition community and as we argue in this report, it is one they seem to be meeting very well.

2. FINGER AND FACE: PERFECT TOGETHER

Finger prints and facial photographs have since long ago been the primary biometrics of choice for the law enforcement community (Moenssens 1969). This could be because they have complementary advantages and strengths. Finger is a precise measure of identity but it is difficult for the untrained human to match efficiently. Face on the other hand interfaces much better with human operators since humans are remarkable in their ability to recognize faces. This makes face ideal for applications where quick verification is necessary (ID cards) or identification from a pool of suspects (mugshots).

In the information age, finger and face continue to be the most significant biometrics in the digital records. The vast majority of law enforcement databases under development either contain one, the other or both. Given the prominent role finger and face play, the question is does the technology exist today that allows efficient and accurate computerized search of finger and facial records?

Some finger print matching technology has been commercial for several years now and today there are incipient standards as well as several choices of technology vendors. While there is room for improvements and further development – especially in the area of matching from "live-scan" devices—It is safe to say that a relatively affordable technology exists today that will meet some of the finger print processing needs of the law enforcement community. For more information see the Biometrics Consortium Home page at http://www.vitro.bloomington.in.us:8080/~BC/.

Computerized face recognition on the other hand is a much more recent addition to the family of biometric technologies but it is rapidly developing to the same level of commercial maturity as finger print. It has some intrinsic advantages over many of the other biometrics that include its non intrusive nature, higher social acceptability, and the fact that human backup is possible when system failure occurs.

One of the leading commercially available face recognition systems is one we were involved in developing and is called FaceIt™. In what follows we will focus on how FaceIt can be integrated into information systems to serve the current needs of the law enforcement community. Those interested in technical information on how FaceIt works can consult the Appendix.

3. WHAT IS FACEIT?

FaceIt is a software recognition engine that was developed by Visionics Corporation. It runs on any Pentium PC under Windows 95/NT and is capable of performing face recognition from live video or from static images. Video input can be through any video capture board such as the Matrox Meteor or the Digital Eyes PCI board or through a digital camera such as the Connectix QuickCam. The static input can be either TIFF or GIF in grayscale or in color.

FaceIt is fully automated in the sense that no user intervention is required. The software locates human heads anywhere in the image, extracts the face and recognizes the person. In recent government testing, the so called FERET test, conducted by the Army Research Laboratory on behalf of DARPA, FaceIt outperformed by a significant margin other competitive face recognition systems from leading research laboratories in the US and Europe in every category tested (Rauss, Phillips, Hamilton and De Persia 1996).

Since that test was conducted in November 1995, the development of FaceIt continued and upcoming releases of the technology will feature even more powerful algorithms and performance.

Currently there are three categories of law enforcement applications that can be built using FaceIt. Those are (1) applications with face as a primary biometric , (2) with face as a secondary biometric and (3) surveillance. Next we describe each in some detail and when possible will give concrete examples of implementations currently underway.

4. FACE AS A PRIMARY BIOMETRIC

4.1 The G.R.E.A.T. Project

In many instances facial photographs may be the only biometric data available in a law enforcement database. This is the case for example in the so called G.R.E.A.T. (Gang Reporting, Evaluation and Tracking) database which was established in 1985 by the Los Angeles County Sheriff's department and was developed since by the Law Enforcement Communication Network (LECN) --- a consortium of more than 650 local police departments. We will discuss this example in some detail since this success story highlights the benefits to the law enforcement community of adopting new technologies and provides a model for how we propose to work with that community to introduce face recognition.

The purpose of the G.R.E.A.T. database was initially twofold: to provide a method for handling the massive number of case files relating to gangs and gang members, and to harness the power of technology and put it to work in processing gang intelligence. The G.R.E.A.T. software features a database search engine capable of searching for keywords. With this system officers were able to move from archaic and difficult methods for searching paper files that number in the thousands, to a new system of automated records that are just a keypress away. With G.R.E.A.T., the process of connecting information to an actual gang member who can be found and questioned has been streamlined. Now, an investigator only has to enter his or her information - no matter how sparse - into the computer and ask it to conduct the search for him or her. All matching records are retrieved and displayed, drastically reducing search time and man-hours and drastically increasing the chances of identification, apprehension and prosecution.

The information contained in the G.R.E.A.T. database comes directly from law enforcement agencies who have signed onto the system. Each agency's computers connect to a regional computer workstation, to which they upload information on local gangs and gang members. This information, in turn, is available to all criminal justice agencies nationwide who choose to participate, providing a large database of value to gang investigators.

Currently, G.R.E.A.T is used by more than 650 local, state and federal law enforcement agencies throughout the nation. This distributed database contains over one quarter of a million records, each record containing 150 possible fields for data entry including vital statistics, gang moniker, tattoos, known affiliates and several reasonable quality facial photographs and more. Using G.R.E.A.T., officers have at their fingertips accurate, constantly updated facts, background, histories, records and photographs of criminal gang, drug, crew and posse group members.

4.2 The Need For Face Recognition Systems

While the G.R.E.A.T. database system allows search by key words, it -- just like all other database software available today -- *cannot* search records by photograph content. Photographs of gang members are routinely taken by officers in the field using portable video cameras. Later at the station, the video footage is digitized by a video capture card on a PC and several images are captured and are attached to the gang member's record in the database. It is important to emphasize that the records contain no other biometric information, such as fingerprint -- collection of this type of information would be very difficult in the field, especially in comparison with the ease with which officers convince gang members to pose for video.

For law enforcement officers attempting to deal with the ever-increasing number of gang members in different jurisdictions, the ability to execute a computerized search by submitting a photograph of an individual can open up whole new avenues for investigations. The need for such capability will only become more severe as the database sizes swell. The current G.R.E.A.T. database contains more than 250,000 records and the number of entries in the system is expected to continue to rise as more local police agencies join the program and upload their gathered entries to the database.

Without computerized face recognition systems, a wealth of information within current databases cannot be accessed by investigators. This would be unfortunate since very often the photographs contain the most reliable information in the record. Gang members routinely present false identity cards and give police false information. It is thus important to be able to identify a gang member through matching a face to images in the computerized records.

Face recognition is also necessary for reconciling the multiple records within a gang database and among different law enforcement databases. Multiple records for the same individual exist in databases since it is not uncommon that records are introduced in different localities by different agencies working with different information. To be able to pool information resources entails the efficient reconciliation of records. A computerized face recognition system is required for achieving this.

4.3 FaceIt DB: A high performance search engine

At Visionics we have designed a search engine called FaceIt DB that meets the requirements of the gang tracking problem and many others like it. This is software that runs under windows 95/NT and can access any database of images in standard formats. It can be queried either by loading a static image in or by using a live video feed. No manual intervention on the part of the operator is required. The software locates the head in the image and extracts it and computes the matching score against entries in the database. It returns to the operator the top 25 (this number can be customized) matching records for visual inspection and determination by the operator (See Figure 1).

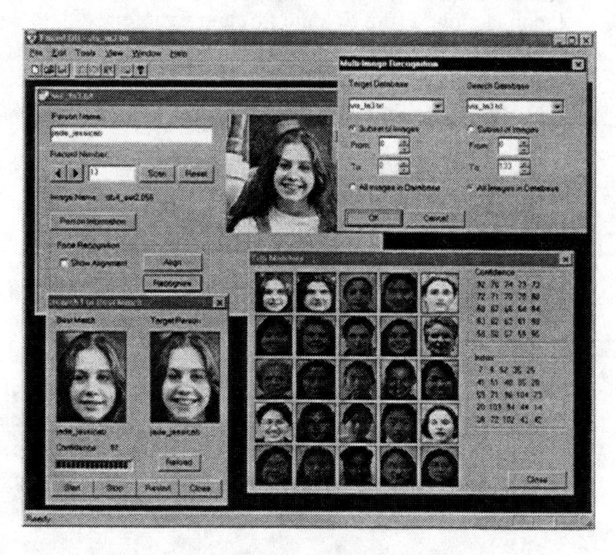

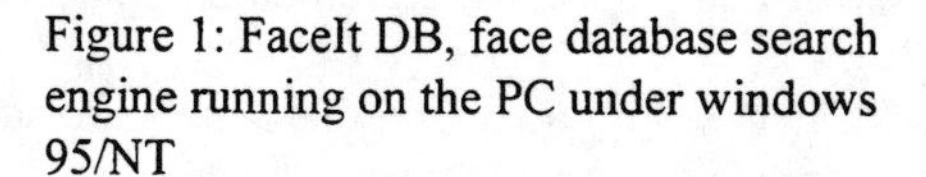

Figure 1: FaceIt DB, face database search engine running on the PC under windows 95/NT

The system currently performs matchings at the rate of 500 per second directly on images coded in standard format (such as TIFF) on a Pentium 100Mhz PC. This speed can go up to more than 50,000 per second if the database is first preprocessed into the internal representation used at the matching stage. FaceIt interfaces seamlessly with the G.R.E.A.T. database engine to give police the added power of face and key word search..

4.4 Internet Version

While FaceIt DB is a very useful stand alone application, its full potential is achieved when it is integrated into a client/server solution where at the server end FaceIt DB and the database engines run, while queries are submitted from client PCs via the internet or an intranet (or a local area network). The database is kept at the server end which is maintained by a regional or national law enforcement consortium such as the LECN or a sworn agency. It could also be kept at a local server within a large police department. The client is any police department with web enabled personal computers (PC) and access privileges to the database. It can also be an officer in the field with a laptop computer with PCMIA 14.4 cellular modem card.

The internet tracking system will work as follows:

- A police officer points the web navigator (Netscape) to the gang server homepage and provides a secure password to gain access.
- The officer is then prompted by the server to fill out a Query Submission Form that asks him or her to provide a photograph of the suspect and any other known information.
- The submitted image and information are processed by the face recognition and database search engines running on the server. Before attempting face matching, the engine locates the face in the submitted photograph, extracts it, transforms it into the internal representation and then proceeds to compute its matching score with photographs in the database. High scores mean high likelihood that the individual in the photograph is the individual in the matching record. Since in large databases it is possible that several people may resemble each other the server will communicate the top 5 matches to the officer for further examination. The officer can double-click on any of the top 5 images to recover the full size image and the record attached to it. He/she also has the option of recovering the next 5 matches etc. if required.

Advantages of the Internet face tracking system:

- **Exceedingly Low Cost**: The system is designed to tap into existing infrastructure such as the internet and does not require any significant new investments on the part of a police department. All a police station needs to get started is a web enabled PC (a 386 or 486 computer is sufficient) with an internet connection ($10-$20 a month) and a subscription to a gang tracking database (price will be determined by police associations depending on number of subscribers. Currently the LECN annual membership fee is $ 250). The hardware at the server end is a collective one time investment and in many cases it could be acquired through corporate donations (e.g. SUN corporation has just donated 50 servers to LECN).
- **No Maintenance or Hidden Costs**: Since the search engines reside at the server there will be no maintenance required at the police end. We envision that the annual membership dues that entitle a police department to access privileges will pay for the server equipment, the search engines and for maintenance and support of the system.
- **Scalable Solution**: The system can be scaled up without difficulty with increasing database sizes and with increasing use by adding additional servers without changing anything at the client end.
- **Never Outdated**: As the technology improves with time the server software can be upgraded easily and all police departments will see the benefit without having to buy any new software.
- **No Training Required**: to learn how to use the system is as simple as surfing the internet, thus no new personnel are required and no special training is necessary. An hour demonstration of the system will be sufficient to enable most officers to use the system.

5. FACE AS A SECONDARY BIOMETRIC

In several other applications, particularly, entitlement programs and enrollment programs such as drivers licenses, passports etc, agencies have the option of using finger images as well as face. Currently several states (e.g. California, Colorado, Florida, Texas) are in the process of establishing biometric screening of drivers to make it difficult for individuals to obtain duplicate licenses under different identities and generally to reduce tampering and faking of licenses. These states already record finger print images in addition to a face photograph. Our informal survey of a dozen DMVs across the country suggests that many states are likely to follow suit shortly. Concurrent with this, is an ongoing effort on the part of a larger number of states to modernize and convert their DMV records into digital form. So the stage is set for the utilization of new biometric technologies by many DMVs around the country. This is also the case in national identification programs in foreign countries, entitlement programs in the US and abroad (welfare in the US, healthcare in Canada), and in other branches of the justice department (INS registration services).

The question that law enforcement officials may have at this stage is how to best take advantage of the two biometric technologies in the short as well as long terms. It is important that any technology adoption plans be flexible enough to allow for the expected advacements in both domains over the next several years. Our recommendation at this stage is to avoid systems that require proprietary hardware or require large scale integration of PC's in local clusters. We believe that an effective biometric system that runs on a handful of inexpensive PCs can be put together today and can be made available through the internet/intranet to serve distributed needs.

The system we are thinking about is one where finger would be used as a primary biometric to generate a candidate list on the order of a 1000 people. By allowing a large candidate list we can benefit from the speed advantage of less complex algorithms and one can go through databases with 5 million records in about a minute on a single PC using commercially available finger print search engines. This candidate list is then fed into a face search engine which in less than 1 second can bring this list down to a few people (or to just one). A human operator can then go through and perform the final determination.

6. SURVEILLANCE

The above are only two examples of how face recognition technology can be used by the law enforcement community. They illustrate its benefits as a problem-solving technology and an efficient investigative and time saving tool. But there are many other ways face recognition can be of great use to this community. For example one can use face recognition to build intelligent surveillance systems. In fact face recognition is the only biometric that is amenable to this function since a facial image can be taken without requiring the participation of the subject. FaceIt, being capable of continuous recognition from live video, is well suited for surveillance applications. It can be constantly on the lookout and will capture and will attempt to recognize any face that comes into its field of view. In the current implementation of the video version of FaceIt, it also keeps a time stamped log of people it has identified and anyone that it does not recognize it keeps their picture in a "strangers" folder.

FaceIt can also be used to control video cameras for more effective surveillance. This is because FaceIt is capable of not only determining the location and size of the human head but it can also track it. This means that FaceIt can be used as a controller on an active vision system where the direction, zoom, pan and tilt of the camera can be adjusted to maintain the subject in optimal field of view.

FaceIt Applicant Processing System

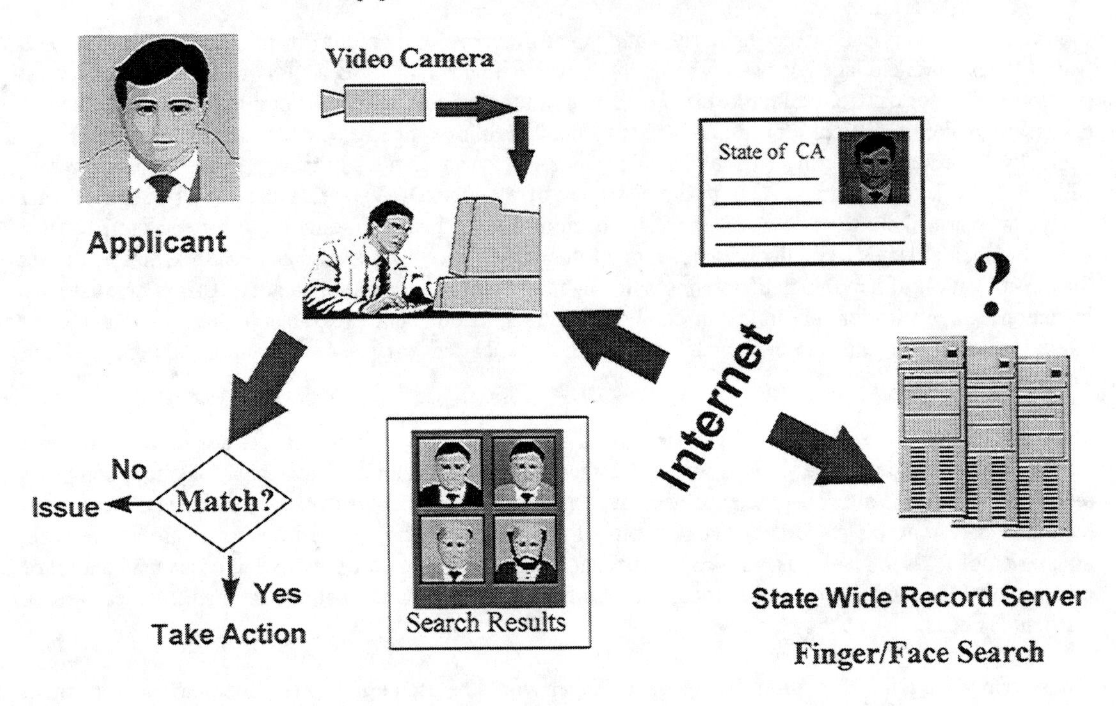

Currently there is an ongoing effort to develop a surveillance system for airports using FaceIt. The need to be on the lookout for terrorists and criminals is very high at the nations airports but also from a technical point of view airports are ideal for surveillance applications since they offer a variety of choke points where a screening system can be placed. A camera installed in the gateway can transmit video to a centralized PC located in secure facility elsewhere at the airport and running FaceIt. This system can screen in real time passengers as they disembark the plane.

Other applications as well as further information on the above can be found at the following websites: **http://www.faceit.com and http://venezia.rockefeller.edu**.

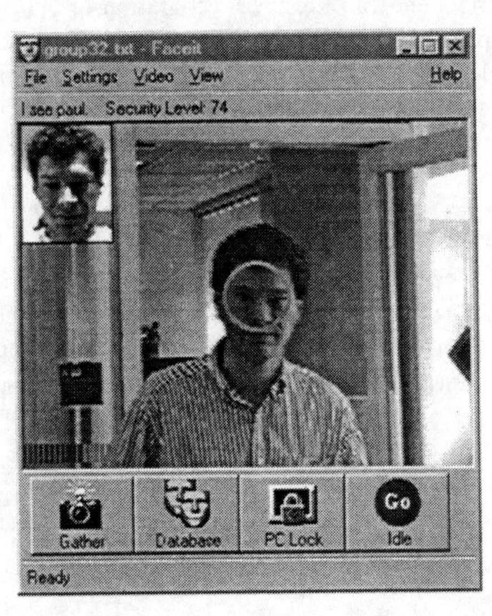

Figure 3: FaceIt can be used as a surveillance system. It automatically picks up any heads in the scene and attempts to recognize them. If never scene before it adds them to its surveillance log.

APPENDIX A: A Technical Primer on FaceIt Technology

FaceIt is an outgrowth of fundamental research at the Institute for Advanced Study on information representation in the late eighties and more recently at the Computational Neuroscience Laboratory at Rockefeller University on understanding how the mammalian visual system processes and perceives complex patterns. Commercial development as well as continued R&D on FaceIt is done by Visionics corporation which was founded in 1994 by the same scientific team responsible for the algorithms underlying FaceIt. In what follows we give a brief outline of some of the key technical features of FaceIt, our discussion will follow the processing stages in FaceIt.

1. Detection: The very first step in face recognition is to find out, in an efficient way, where the face is in the field of view. In a live system, this entire process should not take more than 50-100 milliseconds or else it would constitute a serious bottleneck . The challenge here comes from the fact that the field of view of a camera is relatively large (640x480 pixels) and in principle one is computing the probability of a face at every location. The solution to this problem is to adopt what we call a multi-scale and multi-cue face finding strategy.

In a multi-scale approach one attempts to locate faces in low resolution first and switches to high resolution analysis only at locations where a head-like signal is detected. In our current system we use three scales (1/4,1/2 and 1). The threshold for detection for each scale is determined by optimizing an appropriate measure on the ROC curve for head detection subject to minimal time constraint. Currently the entire face detection component of FaceIt works in less than one tenth of a second on standard PCs and it can detect faces with high accuracy all the way down to faces where the inter-eye distance is about 24 pixels.

Multi-scale is one way to reduce the search dimensionality and it is a strategy used by the mammalian visual system (Van Essen and Anderson 1987) but it is certainly not sufficient for a real time system. Luckily faces on a background produce strong cues that can simplify the search. The most prominent are discontinuities in the spatial, temporal and color domain. One way to reduce the dimensionality of the original search problem is to process the input video/image using the appropriate spatial, temporal and chromatic derivatives and to examine further only those regions where the discontinuities are significant. The precise form of these filters as well as the thresholds can be statistically derived from an ensemble of video segments and images of faces on a background. These cues can also be processed in parallel and in separate threads and only when available. For example when the image is static or the camera is black and white the system relies only on the spatial cues only.

2. Alignment: The problem of ascertaining the existence of a head is not as difficult as precisely determining its position, size and pose. This requires detailed shape and feature detection. For this purpose, our system, in regions where the detection module has given high probability of a head, integrates the edges into contour segments and template matches them against prototypical contour shapes of heads. If this exceeds a given threshold the signal is passed to the next stage which performs facial feature detection in order to verify that this is indeed a face and in order to determine its alignment parameters. Currently we rely on detecting the eyes and the nose. If these features exist then the target is a face and is passed on for further processing as we describe next.

Precise alignment is a significant key to successful face recognition systems. This is because variability due to alignment errors dramatically affects the matching score of a face with the stored templates and leads to false entropy. FaceIt estimates the alignment parameters (position, size, rotation and to a lesser degree pose) by reconciling three different types of information for added robustness. FaceIt uses the contour radius of curvature, the inter-eye distance and also a 3D model of the human head (Atick, Griffin and Redlich 1996) to find an independent estimate of those parameters.

3. Normalization: Once the parameters are estimated, the head is normalized by scaling, rotating and warping. Actually one also needs to normalize with respect to lighting variability. For this we adopt two strategies. We compensate for large scale lighting variations on the face by first estimating the lighting pattern by relying on 3D model of the face but we also perform a 2D transform that we can prove creates a good degree of light invariance. This transform is a byproduct of our work on shape-from-shading (Atick, Griffin and Redlich 1996) and it allows us to construct the same output irrespective of how lighting is changed over the face, of course within a reasonable range. The details of this transform remain proprietary to Visionics.

4. A face print: Ultimately every face recognition system has an internal representation for faces. This is a coded version of the face that is better suited for matching against the database. In recent years the problem of finding computationally viable representations for complex objects has attracted a lot of attention within the pattern recognition and applied mathematics community. The outcome of this is a dozen or so methods that allow the extraction of a representation from an ensemble of examples, these can be categorized as statistical, neural network and AI methods. Here we will examine two simple yet powerful techniques that have been used in face recognition, one is Principal Component Analysis (PCA) and the other is Local Feature Analysis (LFA).

The application of PCA to the problem of face representation was first done by Sirovich and Kirby in 1987 and they coined the term eigenfaces for the global eigenfunctions that form the basis of their representation. Later they were used by Turk and Pentland and many others in different face recognition systems. It is now clear, however, that eigenfaces are not the best representation for faces and that they suffer from severe limitations. For example they produce global nontopographic features – which is undesirable as we shall see next -- and they are not as low-dimensional as one may think a priori if one is to use them to represent a diverse group of people in diverse lighting conditions.

In Figure 4 we show some eigenfaces that were derived by diagonalizing the covariance matrix of the images from the FERET database (Rauss et al 1996). As one can see many of the most significant modes are not even face modes but they capture entropy due to lighting variability. One can also see that they are global which means they cannot be invariant with respect to local variability in the face coming from expression changes or speech. The eigenface representation is also nontopographic, which means that phase relationships within a face (position of eye relative to a nose etc) are not preserved by the representation. This is a severe limitation for representing complex patterns.

In addition eigenfaces provide a linear approximation to face space since they represent a face as a linear sum of modes about a mean face. This is an excessive approximation of a space that can be shown to be nonlinear in cluster analysis. Of course one acceptable remedy for this is to linearize about several cluster centers in face space, but this is not what is done, as far as we can tell, in current eigenface based recognition systems.

The advantage of eigenfaces, of course, is their computational simplicity. Thus one can ask if a representation--as simple as eigenfaces but without their limitations —can be derived? The answer is yes and it comes from LFA.

LFA is a mathematical technique for deriving local, topographic representations for a class of complex objects from an ensemble of examples. It was first introduced in vision by Atick and Redlich (1990) and was more recently formulated as a general mathematical theory for object representation by Penev and Atick (1996). Here we will give a very brief description.

In LFA we start by diagonalizing the covariance matrix

$$R(\mathbf{x},\mathbf{y}) = \langle \varphi(\mathbf{x})\varphi(\mathbf{y}) \rangle$$

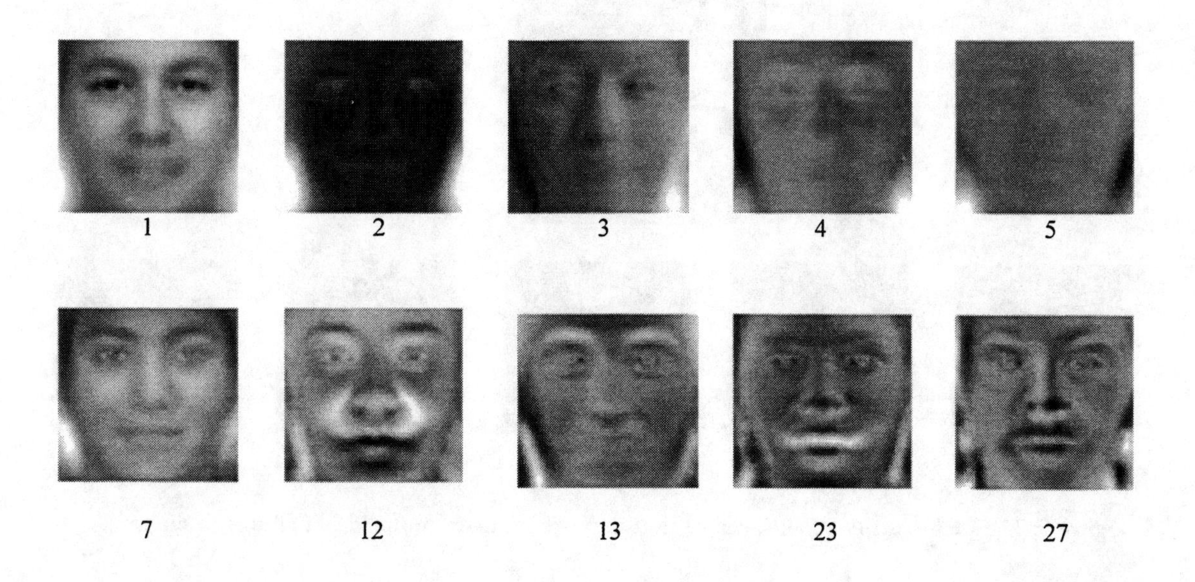

Figure 4: A sample of eigenfaces. Distadvantages are apparent in their global nature and their strong dependence on light.

where $\varphi(\mathbf{x})$ is the face image and < > denotes ensemble averaging. This assumes that faces have been properly aligned by some mechanism. One convenient alignment method is to align by the eyes. The diagonalization produces eigenmodes $\psi_n(\mathbf{x})$ with associated eigenvalues λ_n, these are the familiar eigenfaces. The covariance matrix for faces is well known to be rank deficient and one can take advantage of that to diagonalize a much smaller matrix with rank given by the number of faces over which the ensemble average is done using a procedure called the snap shot method first used by Sirovich and Kirby (1987).

LFA constructs the following topographic kernels

$$K(\mathbf{x},\mathbf{y}) = \sum_{n=1}^{N} \psi_n(\mathbf{x}) \frac{1}{\sqrt{\lambda_n}} \psi_n(\mathbf{y})$$

where N is the rank of the matrix R. These kernels can be proven to give the most decorrelated output (hence efficient):

$$O(\mathbf{x}) = \int K(\mathbf{x},\mathbf{y}) \varphi(\mathbf{y})$$

but more interestingly they lead to a representation in terms of local features. In Figure 5 we show the kernels derived from the ensemble of faces in the FERET database. As we can see, the receptive fields develop compact support and are local. They are also strongly matched to the features of the face. For example, a receptive field matched to a mouth develops at position (a), a nose receptive field—at position (b), and eyebrow, jawline and cheek-bone receptive fields—at positions c,d, and e, respectively. Note that the receptive fields that develop are not edge detectors in general; they are feature detectors, different from each other, and matched to the feature that is expected near their respective centers.

The LFA is a representation, in the sense that the original image can be reconstructed from the LFA output without significant rms error. In Figure 6 we show an out-of-sample image, the reconstructed image and the error. The representation is strongly identity preserving. Actually what is more interesting about it is the

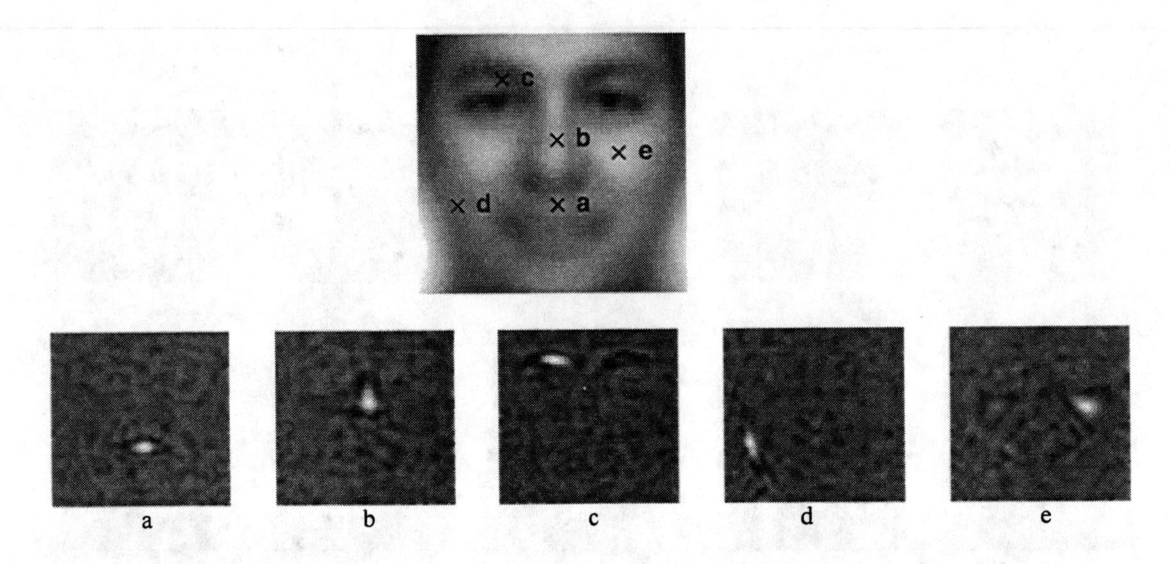

Figure 5: The LFA receptive fields. Notice that they are matched to the local features of the face.

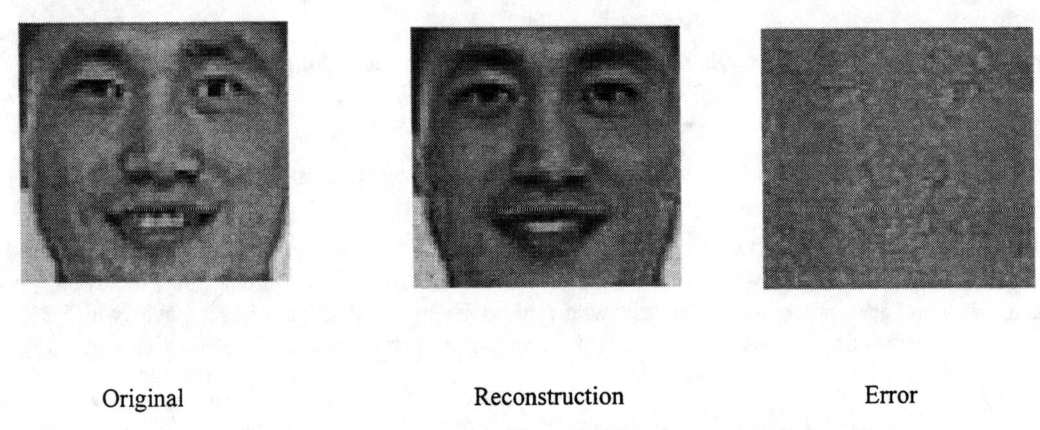

Figure 6: Reconstruction with the LFA Representation

fact that it can be sparsified to produce a low dimensional representation. This is because the outputs possess residual correlations given by

$$< O(\mathbf{x})O(\mathbf{y}) >= \sum_n \psi_n(\mathbf{x})\psi_n(\mathbf{y}) = P(\mathbf{x},\mathbf{y})$$

The existence of these residual correlations means that not all output units are needed for the representation. One can predict the entire output from knowledge of the output at a much smaller subset of points (on the order of 40-64) and from knowledge of the a priori correlation matrix $P(\mathbf{x},\mathbf{y})$. The subset of points that optimally reconstructs the input changes from one face to another providing very valuable information. There is an efficient algorithm that picks these optimal points. It allocates representation resources in regions in the face that deviate from expectation as measured by the metric $P(\mathbf{x},\mathbf{y})$ (for details see Penev and Atick 1996). In Figure 7 we show the first 25 points that are picked by this algorithm.

5. Matching Measure: After a face is mapped into the internal representation or the face print the final stage is to perform matching against the prints stored in the database to determine identity. For this purpose a metric or measure on face space is required. In some ways one can argue that this measure is even more fundamental than the representation itself. A representation may be faithful in the sense that it can reconstruct the original image of the face accurately but what is more relevant is the ability of thc representation to be in one to one correspondence with the identity of the person. As we have seen above in LFA one uses prior knowledge of faces – captured by $P(\mathbf{x},\mathbf{y})$ – to create a representation. It turns out that $P(\mathbf{x},\mathbf{y})$ can also define some powerful metrics on face space. For example let us assume that the face print stored in the database is given by the set $\{O_m\}$ while the probe image when processed leads to the set $\{O_n\}$ which in general may include different set of points. One possible measure is the normalized correlator between the two sets with $P(\mathbf{x},\mathbf{y})$ as the metric, i.e. $E = \sum_{n,m} O_m O_n P(\mathbf{x}_n, \mathbf{x}_m)$.

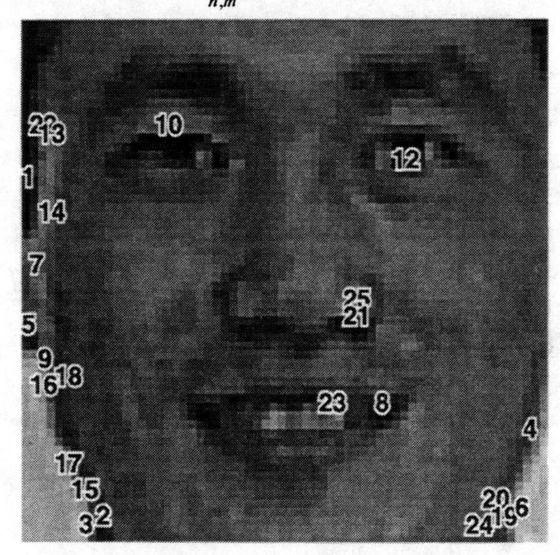

Figure 7: A sparse set of points is enough to reconstruct the whole face

REFERENCES

Atick, J. J. and Redlich, A. N. 1990. Towards a theory of early visual processing. Neural Computation, 2, 308-320.

Atick, J.J., Griffin, P.A. and Redlich, A. N. 1996. Statistical approach to shape-from-shading: reconstructing 3D face surfaces from single 2D images. Neural Computation,8,1321-1340.

Moenssens, A. A. 1969. 'Fingerprints and the law' Chilton, Philadelphia.

Penev, P.S. and Atick, J.J. 1996. Local feature analysis: a general statistical theory for object representation. Network: Computation in Neural Systems., 7,477-500.

Rauss, P.J., Phillips, P.J., Hamilton, M.K. and DePersia, A.T. 1996. FERET (Face-Recognition Technology) recognition algorithms. In Proceedings of ATRWG Science and Technology Conference, July 1996.

Sirovich, L. and Kirby, M. 1987. Low-dimensional procedure for the characterization of human faces. J. Optical Society of America, 4, 519-524.

Turk, M. and Pentland, A. 1991. Eigenfaces for recognition. Journal of Cognitive Neuroscience, 3,71-86.

Van Essen , D.C. and Anderson, C.C. 1990. Information processing strategies and pathways in the primate retina and visual cortex. In: Introduction to Neural and Electronic Networks, ed S.F Zotnetzer, J.L. Davis and C. Lau. Academic Press, Orlando, FL.

Error Diffusion Techniques for Printing Fingerprint Images

Harsha Wabgaonkar
Mehron Vaezi
Behnam Bavarian

Printrak International
1250 N. Tustin Ave, Anaheim CA 92807

ABSTRACT

One of the common techniques of printing gray scale images using a bi-level device such as a commercial laser printer is Error Diffusion. Since most of the fingerprint images are gray scale images, they too can be printed using error diffusion. However, none of the existing popular error diffusion algorithms makes use of the ridge-flow information present in a fingerprint image. In this note, we explore the use of ridge-flow directions in a fingerprint image for the purpose of the error diffusion in printing fingerprint images.

Keywords: Printing, Fingerprints, Error Diffusion, Halftoning

INTRODUCTION

Fingerprint impressions are obtained by various means such as by inking and then rolling the fingers over paper, or by live-scanning (similar to photo-copying), or by "lifting the impressions" from a scene of crime, etc. These impressions are eventually converted to an electronic image form, which at some point or the other, need to printed for the purpose of presenting evidence, maintaining records, or for rescanning of the printouts for further processing, etc. The electronic images tend to be gray scale images, typically with 256 gray levels at a resolution of 500 dots per inch (dpi). Most of the applications demand that these multi-level images be printed using an off-the-shelf commercial laser printer, which is strictly a bi-level device. The problem of printing multi-level images on binary devices is a classical one, and had received considerable attention over years([1]-[3]). The problem area is generally referred to as "halftoning". With the advent of low-cost high resolution display devices, high bandwidth communication links, and low-cost high resolution laser printers, this area has once again attracted the attention of many researchers (e.g., [4], [5]).

One of the most popular techniques of image halftoning is Error Diffusion. In this technique, the gray scale intensity of a given pixel is first quantized based on the choice of the printer (raster) resolution, resulting in a quantization error at each of the pixels. The quantized levels are then mapped to patterns of ink dots, so that the entire image is then represented by a matrix of ink dots that the printer prints out on the paper medium. Error diffusion refers to the process of distributing the quantization error at a given pixel

SPIE Vol. 2932 • 0-8194-2334-3/97/$10.00

to its neighboring pixels. Generally, the closer the neighbor spatially, the higher is the amount of the quantization error feedback it receives. Specifically, if p is a pixel in the input gray scale fingerprint image at row r, and column c, and if $I(p) = I[r][c]$ is the intensity of the pixel p, then the quantization error is given by

$$e = I[r][c] - Q(I[r][c]),$$

where Q() is the quantization function. It is usually characterized by the choice of a threshold *T*, which itself could be a single value, or a set of values (multi-level thresholding). If p_N is any pixel in the neighborhood *N()* of p, at the location [r+i][c+j], then, after error diffusion, the new intensity at p would be:

$$I(p_N) = I[r+i][c+j] + w[i][j]*e,$$

where w[][] is the error feedback weight. The pixel p_N may itself belong to many other pixels' neighborhood. Therefore, the final intensity at p_N is the sum total of the net feedback and the pixel's original intensity value. Implicit in the technique is the fact that, the pixel p will be visited and processed first, and at some later instant of time, the neighbor pixel p_N will be visited and processed, so that causality is maintained in a single-pass algorithm. Thus, the error diffusion process can essentially be specified by the following three parameters:

1. The threshold *T* used in the quantization process
2. The choice of the neighborhood over which the quantization error needs to be distributed
3. The amount of the quantization error feedback that each neighbor pixel must be receive. This is usually identified as the corresponding "weight" of the error filter.

All of the popular error diffusion algorithms in the literature emphasize:

1. Simplicity of computation
2. Causality: Specifically, distribute the error to those neighbors that are yet to be visited, so that a one-pass raster-mode operation is feasible
3. Pleasing appearance of the printout to the human eye

They, however suffer from many problems, such as, the presence of various artifacts, usually a low-pass (or a band-pass) behavior, transients at the extremities, etc. More importantly, to us, they are generic, and hence do not take the quintessential ridge-valley structure of a fingerprint into account. Also, most of the algorithms stress the pleasantness of the printout to the human eye. On the other hand, the fingerprint printout is usually examined by an expert looking for certain domain-specific features, or is used for scanning the image using an electronic scanner. In this note, we present a very brief outline of our work in which we have tried to incorporate the ridge-flow direction information in error diffusion. It is described in the next section.

APPROACH

Many sophisticated algorithms exist to extract the ridge-flow direction information from a fingerprint image. Typically, this information is produced as a by-product when a fingerprint image is processed for extracting minutiae and other features. It is represented as a ridge-flow map d of a fingerprint image, in which d[r][c] represents the direction of the ridge-flow at the given location (r, c) in the fingerprint image. The direction could be represented by an angle theta, or by an indirect representation of the angle, e.g., sin(theta), cos(theta), etc. The particular algorithm we have used produces two types of values in the direction map:

1. No Direction: For example, if a pixel lies in a blank (white) space or in a dark ink blob, where no valid ridge-flow exists.
2. Valid Direction: For example, in terms of theta, these values could be between 0 and 360 degrees.

Our error diffusion algorithm is as follows:

> if a pixel direction has a "No Direction value"
>
> use the *Floyd-Steinberg Error Diffusion Algorithm*
>
> *else*
>
> diffuse the error along the pixel direction

The Floyd-Steinberg Error Diffusion algorithm prescribes the following amounts to be distributed to the neighbors: Generally, as stated earlier, the closer the neighbor distance-wise, the higher is the amount of the quantization error it receives.(Assume that row-coordinate increases in the downward direction, and the column-coordinate increases to the right), the algorithm is as follows:

If e is the quantization error at a pixel with row and column coordinates (r, c),

- (7/16)*e is given to the neighbor on the immediate right, i.e. one with coordinates (r, c+1);
- (1/16)*e is given to the neighbor on the lower right diagonal side, i.e. one with coordinates (r+1, c+1);
- (5/16)*e is given to the neighbor immediately down below, i.e. one with coordinates (r+1, c);
- (3/16)*e is given to the neighbor on the lower left diagonal side, i.e. one with coordinates (r+1, c-1).

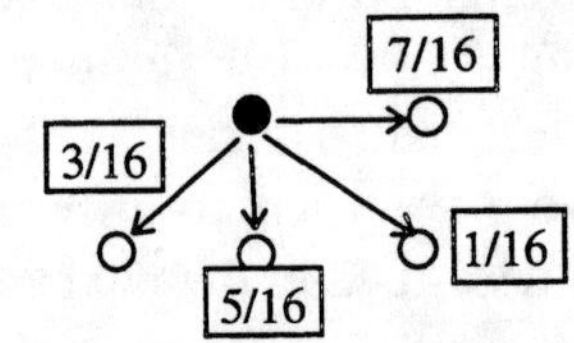

The Floyd-Steinberg Error Diffusion algorithm

If a pixel has a valid direction, we use the neighbors of the anchor pixel along that direction for error diffusion. In the figure below, we have shown six neighbors of the anchor pixel (shown dark) along approx. 45 degrees. The error diffusion is done as follows:

- (1/2) * e to the immediate down neighbor;
- (1/4) * e to the remaining two neighbors.

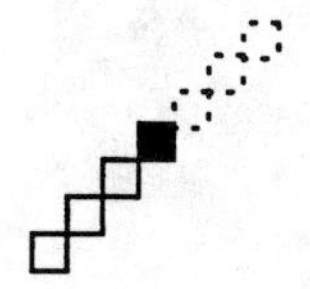

An anchor pixel (dark) with 3 neighbors on either sides along 45°. Only the causal ones participate in error diffusion.

Note that the algorithm is causal (i.e., the error is distributed only to the neighbors that are to the right and down below the anchor pixel). This algorithm is thus a single-pass algorithm. If sufficient resources exist, a non-causal extension of the algorithm is straight forward. However, this requires several passes through the image. The iterative process is stopped when the changes in the intensity become insignificant in accordance with a certain threshold. It is more effective to diffuse error not only along the pixel direction, but in the orthogonal direction as well. Of course, the sign of the error feedback in the orthogonal direction needs to be reversed.

A fingerprint image (800 x 750 at 500 dpi) was printed at 1200 dpi on a commercial off-the-shelf laser printer. The printer output for the causal algorithm for various magnitudes of the error feedback is shown below. The first row (a) shows the printout without using the direction information, and is based entirely on the Floyd-Steinberg Error Diffusion algorithm. The second row (b) row shows the output of the combination algorithm described above, which incorporates both the Floyd-Steinberg Error Diffusion algorithm for no-direction pixels, and directional error diffusion for the rest of the pixels.

(a) Only Floyd-Steinberg Error Diffusion algorithm:

Feedback: e

Feedback: 4*e

Feedback: 16*e

(b) Combination Algorithm

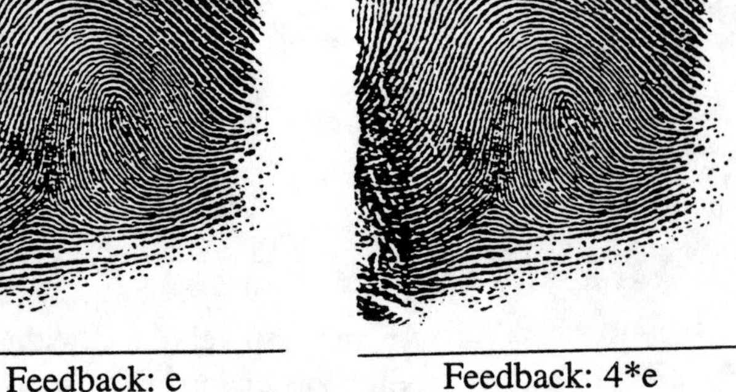

Feedback: e Feedback: 4*e Feedback: 16*e

CONCLUSION

Traditional error diffusion algorithms being generic, do not incorporate the fingerprint ridge-flow direction information for halftoning. In this note, we have presented a brief outline of our work that does attempt to exploit the ridge-valley structure in improving the quality of fingerprint hard copies. Further research work will try to characterize the frequency response of our algorithm, and its termination criteria in a multi-pass version.

REFERENCES

Floyd, R. W., and Steinberg L., "Adaptive Algorithm for Spatial Grey Scale", Proc. SID, 1976, vol. 17/2, pp. 75-77, 1976.

J. F., Judice C. N., Ninke W. H., "A Survey of Techniques for the Display of Continuous-tone Pictures on Bilevel Displays", J. Electron. Imaging, vol. 2, no. 3, pp. 193-204, July 1993.

Ulichney R. , Digital Halftoning, Cambridge MA, MIT Press, 1987.

Crounse K. R., Roska T., Chua L. O., "Image Halftoning with Cellular Neural Networks", IEEE Trans. Circuits and Systems-II, vol. 40, no. 4, pp. 267-283, April 1993.

Zakhor A., Lin S., Esakfi F., "A New Class of B/W Halftoning Algorithms", IEEE Trans Image Proc. vol. 2, no. 4, pp. 499-509, October 1993.

Addendum

The following paper was announced for publication in this proceedings but has been withdrawn or is unavailable.

[2932-19] **Real-time non-contact 3D finger print input system**
S. Yin, H. Liu, The Pennsylvania State Univ.; S. Jutamulia, Photonics Research; T. Lu, Physical Optics Corp.

Author Index